AF303574

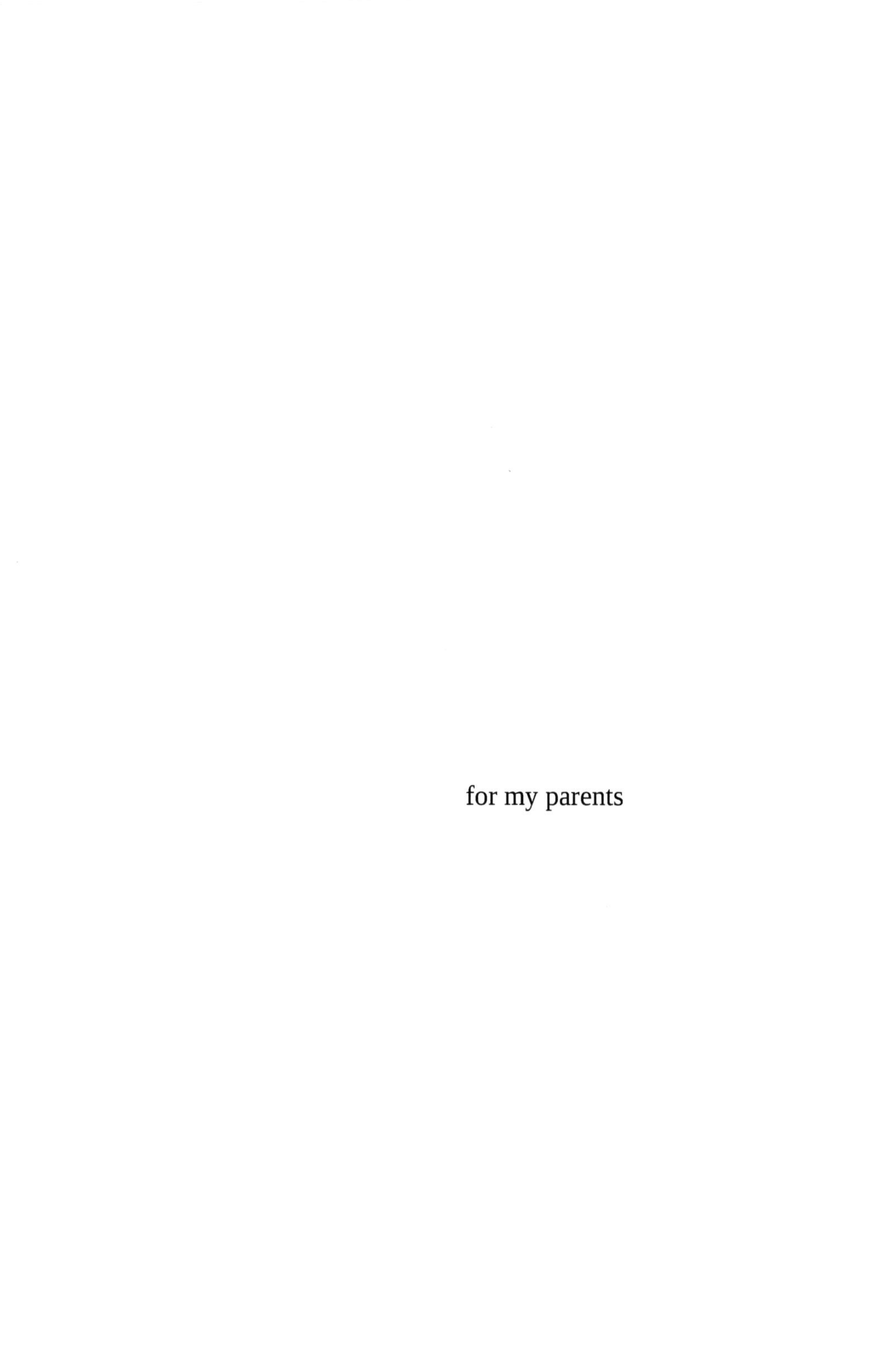

for my parents

# The Relativity of the Observer and the Gravity

# Table of Contents

# Introduction

There are more and more books in which increasing problems in physics are described. Actually feasible observations are no longer compatible with each other. Sabine Hossenfelder[13] writes in her book: "... *Looked at in another way, rigidity could also mean that we have gotten into a dead end, that we have to rethink problems that have long been solved and have to look for a path that has not yet been taken.*" (translated by me)

If the authors are competent in the field, they can make the problems clear. But it also shows that those who make the decisions do not dare to actually question the dogmas of modern physics. But without questioning the basic assumptions, we can no longer get out of the impasse. This book is written for people who are interested in physics and are willing to look at such dogmas from a different angle.

You don't have to read the book in chronological order. You can choose individual chapters. This is not a textbook. The basics of physics and the theory of relativity are more likely to be found in such. I have described here more in a philosophical sense what position man has in the universe that surrounds us and what the consequences are. The reader interested in relativity theory can also start straight away with the chapters on Minkowski diagrams and the method of measuring the motion to the gravitational field. Although he might also be interested in the chapters on space and time.

Einstein once said that one must question dogmas in order to break their too great power. (translated by me)

My teachers never really questioned anything. They only asked a question, only to then lay out tracks along which the question gets the answer the teacher wants.

Far from the tracks, this book deals with these questions in a very unconventional way. Observations are interpreted completely differently and solutions are then offered without placing too much emphasis on mathematics. Mathematics (see Chapter 1.3) can represent anything you want, as if it were a language. This leads to multiverses that may all be mathematically equivalent. In this book, only the universe that really surrounds us is to be examined. What is meant by "real" is described in chapter 1.4 on reality.

All ideas are based on causality, linearity and a self-contained logic. The reader should be challenged to find errors in this logic. Or if he can't do that, presenting the ideas to other people who seem competent to him.

More and more phenomena are being discovered that are not explained well or at all, such as the fly-by anomaly [1] or galaxies rotating too fast for their mass. For the galaxies one looks for dark matter [2][25], a strange phenomenon that shows up nowhere except in the formulas to match the actual observations. The experts keep looking for phenomena that could fill the gaps in formulas. What has now become a real industry and is also protected accordingly. Others who doubt the correctness of these formulas have mostly not understood them. In particular, the special theory of relativity is repeatedly called into question. Some construct attempts to refute the special theory of relativity. I have only found examples whose description directly or indirectly contradicts the logical consequences that result from the Lorentz transformations. However, I believe that these correctly describe the relative conditions in the universe around us. I agree with the experts that it is always difficult to uncover the mistakes of the doubters, especially if they are resistant to advice.

Sabine Hossenfelder writes: *"An enormous amount of effort was put into these ultimately failed attempts to find new laws of nature. There hasn't been any progress in basic physics for more than thirty years now.*"(translated by me) Shouldn't the current situation awaken the willingness to question fundamental dogmas.

The question is: what is the basis for the Lorentz transformations? On the basis that the speed of light **is** the same in all inertial systems, in the sense of a natural constant from which other things can be derived. Or that the speed of light is always measured constantly with light clocks under Einstein's definition of simultaneity. This is also possible on the basis of a real field to which the principle of motion can be applied, in contrast to what Einstein expressed in his speech on May 5, 1920 at the Reich University in Leiden. This is further elaborated in Chapter 5 on gravitation.

Ernst Mach changed the reference to absolute space through the reference to the masses in the universe. Einstein didn't shy away from using the term gravitational ether, but he didn't want to include the concept of motion in it. This was required to further maintain the validity of special relativity as a border case and the speed of light as an actual constant to the observer.

On the other hand, one accepts the motion of gravitational waves. However, these can only propagate through a medium to which an observer can then also move. More on this under Chapter 3.8 Wave-particle duality.

There is the wonderful gauge of satellite navigation. Here there is a preferred system of rest, which does not rotate with respect to the starry sky. This also clearly determines the motion of the clocks. If clock B runs slower, from the point of view of an observer A to his clock A, then clock B also runs slower from the point of view of the observer B. All clocks move within the framework of the Lorentz transformation, but in relation to a preferred rest system, as Lorentz himself thought. Is it really different with linear motion?

My question is: Does the general theory of relativity really need the validity of the absolute constancy of the speed of light, or would it not be able to cope much better with the explanation of the observable phenomena without this restriction.

I see mathematics as a language with which nature can be described.[1] However, it has no more to do with nature than a picture description

---

1   Prof. J. Behrens, Universität HH: „Mathematik als Sprache" unter www.min.uni-hamburg.de
    [13] S. 21: … eignen sich nich alle Disziplinen für die mathematische Modellierung – die Verwendung einer so exakten Sprache ist nicht sinnvoll, …
    [18] S. 211: … Man sagt auch in der Sprache der Quantentheorie, …

with a painter's picture, which only reflects what the author has seen in the previous viewing of the picture and determined by examining its nature. There may be hints of further knowledge, but never anything the author has not yet discovered or totally contradicts his views. Reliable predictions about statements not yet made in the description of the picture, about the exact origin or condition of the picture cannot be made from this. The notes contained in the image description must also be checked to see whether they actually exist. By this I mean, for example, a prediction that the analysis of a colour in an area of the image that has not yet been examined would yield the same result. The Nebra Sky Disc[2] shows that one must examine each piece of gold trim to get a better understanding of its formation. The generalization of the first gold sample to the entire gold setting would have led to a false statement here. The disc has been reworked several times and the gold trim has a different composition of gold in different places. Several artists were involved.

When measuring the rotation speed of the galaxies, one can only state that the formula based description and the observed phenomenon of the too fast rotation do not fit together. When discrepancies arise, people tend to stick with what they have learned so far and look for the error elsewhere. This may be successful, as in the case of the Pioneer anomaly[20], whose solution lies in an observation gap. The radiation of energy is more asymmetrical than originally thought. In this case, only the meticulous examination of all individual parts and their interpretation led to success. Postulating dark matter may solve the problem of galaxies rotating too fast without having to challenge the fundamental ideas. But shouldn't the magnitude of this problem prompt a willingness to reconsider the fundamentals of gravity? Does it really need the dogma of the absolute constancy of the speed of light as a prerequisite?

Much of the book is written without formulas and intended more for the philosophical view of the world for anyone to ponder. If you can understand the introduction, you will be able to develop new ideas with most of the book. The chapters with the Minkowski-diagrams require some graphic understanding, but there are no formulas either. This book does not deal with mathematical derivations, but tries to make understandable what is behind the formulas. Mathematical derivations are important in order not to make logical mistakes when drawing conclusions. In principle, however, mathematics can represent

---

2    You can read more about it on the Planetarium Hamburg website.

everything. If one uses cubic instead of square in the case law, the formulas will not be wrong, only they will definitely describe something other than the falling that actually occurs in our universe. In order to be able to work with the formulas, measured values must be used. If our idea of the linking of the measured values with these formulas is wrong, we will only pass on this error logically correctly with the formula derivations. We will not be able to uncover the error as such. To make this clear, I would like to present a formula here:

$2 + 2 = 3 + 1 = 4$, probably a problem that can be solved mathematically without any problems. The problems in physics today that I want to address are not about mathematical processing. It is about measured values, how they are obtained and what is behind them. Symbolically for this problem I would like to ask the question:

Are 2 apples + 2 pears = 3 oranges + 1 lemon? Try as we may, we will not be able to solve this problem mathematically. The question is, what are apples and pears? In German they say "compare apples with pears" (equivalent translation). In the present case, it is obviously the case that things are equated that are not the same. But when we get into the real of relativity, we are dealing with the comparison of 1 m and 1 s that A measures and 1 m and 1 s that B measures. Here the question is: Is 1 m = 1 m and 1 s = 1 s. Here it is much more difficult to see if you are comparing apples to oranges.

The special theory of relativity is undoubtedly correct. Anyone who doubts this has not yet properly understood it. The question is whether the assumptions (Einstein's postulates), which are absolutely necessary for the special theory of relativity, actually apply to the reality of the universe around us? No new formulas are needed for my ideas because I assume that the relative relationships are correctly described by the Lorentz transformations or the Maxwell-Lorentz equations. And the formulas of the general theory of relativity are also correct in assigning the measured values.

Only the interpretation would have to change. For example, on what basis the measured values for meters and seconds are obtained and how the necessary measuring instruments behave during the entire measurement process to be compared.

I would like to quote Einstein's speech on May 5, 1920 at the Reichs-Universität in Leiden: "*In summary, we can say: According to the general theory of relativity, space is equipped with physical qualities; in this sense, there is an Ether According to the General Theory of Relativity, a space without ether is unthinkable, because in such a space there would not only be no propagation of light, but also no possibility of the existence of scales and clocks, thus also no spatio-temporal distances in the sense of physics. This ether, however, is not allowed be thought to be endowed with the property characteristic of ponderable media1, to consist of parts traceable through time; the concept of motion must not be applied to it."*(translated by me)

However, what is theoretically not allowed to be done is often possible in practice. Today it is said: Whenever gravity comes into play, the special theory of relativity does not apply. Could it be because, unnoticed, during the transition from the special theory of relativity to the general theory of relativity, a real field was reintroduced to which the concept of motion can be applied? Some describe the effect of gravity with a **rubber mat** in which masses of different sizes form craters of different depths, thereby deflecting the motion of small masses rolling over the mat. You can stick needles into the rubber mat and no matter how the balls roll, the needles stay in the same place. This rubber mat represents a field to which the concept of motion can be applied, both for rotation and for linear motion.

In the following, I will first present the basic position of an observer. What is reality for him? What is space and time? In the following, I will show at some points why the previous description of gravitation leaves questions unanswered and why the microwave background radiation is of no real help in clarifying the question of whether the principle of motion can be applied to the gravitational field. Then I added a chapter with formulas. You don't have to understand these formulas, you can also skip the formulas and just read the text to understand the meaning. In this chapter I describe a method to measure the motion to the gravitational field. The clock required for this is still missing, but it is also described how you could possibly find it.

In Chapter 4, problems of the theory of relativity are explained using Minkowski diagrams, such as the twin paradox Chap. 4.6 or the question: How do two observers, passing each other, see a flash after one second that they emitted when they met. Cape. 4.2 With a bit of graphic understanding you will be able to understand this well.

If the reader has problems with terms such as Michelson-Morley experiment, Sagnac effect, Schrödinger's cat or the double-slit experiment, he can look them up on Wikipedia. In this book, the explanations would only detract from the actual topic and I couldn't explain it any better than Wikipedia does.

For me the most brilliant thinkers are Darwin and Einstein. They did not blindly follow their teachers and adopt the dogmas that prevailed when they were growing up.

There are many brilliant thinkers who have developed amazing things. However, these have mostly only further developed existing ones or put them together anew. Or they've translated whole new discoveries that the rest of the world hasn't even thought about into understandable and communicable forms. Darwin and Einstein questioned the dogmas of their time and developed completely new principles from them, the scope of which only slowly expanded.

Above all, Einstein continued to be critical of his own ideas to see whether they would continue to fit the new findings. I would like to quote two statements from Einstein:

"That where our calculations fail, we call coincidence."

And from a reply he wrote to M. Solovine in 1949:

*"I am quite touched by your heartfelt letter, which contrasts so much with the countless other letters that have rained down on me on this unfortunate occasion. As you put it, I look back on my life's work with quiet satisfaction. But it's very different seen up close. There isn't a single concept I'm sure will hold up, and I feel unsure if I'm even on the right track. But my contemporaries saw me as both a heretic and a reactionary who, so to speak, outlived myself. This has to do with fashion and short-sightedness, but the feeling of inadequacy comes from within. Well --- I guess it can't be otherwise if you're critical and honest, and humour and humility keep you balanced in spite of outside influences."* (translated by me)


# 1. Reality, belief and knowledge

## *1.1 Basics*

Philosophy offers a huge range of thoughts on this subject. I would like to confine myself to those that are important in this context and that I take for granted. It is important to be clear about what basic ideas and assumptions you are starting from when you are thinking. The moment you take something for granted, you exclude outcomes that are logically incompatible with it.

I believe that I am part of the real universe that surrounds us, and my goal is to explore this universe that surrounds us. The universe that

existed before humans thought about it and that will continue to exist when humans die out and nobody cares about it any more. But there are the questions:
- Who am I?
- Is this universe real?
- And what does it mean that I think about it?
I divide reality into three areas.

## *1.2 The reality of the ego*

The one reality is in the centre: the reality of man. This culminates in the sentence: "I think, therefore I am."[3] But am I the only thinking being? If you know the film "Matrix", it could be helpful to better understand the question. In the film, the people live preserved and connected to a computer in an illusory world, but they are all still individuals. From my personal point of view, however, there is still the question: Is there any other thinking being apart from me? I don't even exist as a human being, maybe there are no human beings at all, and does the universe that I think I perceive exist at all? Bookcases have already been filled on this question, so I'll limit myself to what I believe in. There is no evidence that my belief is correct. I believe that I am just one individual among many, and in this reality as a human being I perceive the environment through my senses and try to understand the world from these sensory impressions.

There is also a very important problem. We humans learn to use our senses in the womb, and this continues after birth. Through these senses we do not perceive nature as it is, but the stimuli that the environment exerts on us are converted into nerve impulses that our brain processes. In this way, I create an image of the environment that does not necessarily have to correspond to it.

This is not only because the physical processes in my head are different from the physical processes I am currently observing. My observation is also strongly influenced by the knowledge I have acquired so far. When we look at the sky today, we no longer see a sun god pulling a golden disc across the sky, but we see ourselves turning with the earth and rotating around the sun. Of course there are also romantic ways of looking at the starry sky. But it is solely through the filter of previously acquired knowledge that an individual can single

---

3    This is the first principle of the philosopher René Descartes (1596-1650) (translated by me)

out the most likely real among the various scientific and romantic possibilities.

At the same time, I always remain the thinking person, for whom knowledge differs from belief only in the repeatability of events. If I misinterpret a process, repeating the event, no matter how often, will not reveal this error. In his book on thinking, Edward de Bono gave an example of logic [5]. In it, a researcher tries to prove that spiders can hear with their legs. He trains the spiders so that they also jump when the command "jump" is given. He then cuts off their **spider legs**, and upon commanding "Jump" again, they no longer do so, as by his logic, they can no longer hear the command.

It will be clear to everyone: the spider needs legs to be able to jump. In the experiment, therefore, a change is made which in any case prevents the jumping motion from occurring, whether the spider hears with the legs or not. For the logical interpretation of an observation it is essential that one has included all necessary parts. If I don't think about the jumping itself, I don't even recognize the logical break in the test procedure.

I think from the reader's point of view all the necessary details of the experimental procedure are clear and hence the error in the researcher's logic. But what happens when the details of a test procedure are not so clear or there are gaps in knowledge despite all efforts? I would like to go into two model examples.

The first I would like to mention is the black powder problem. Imagine being in the Middle Ages and having to explain causally why black powder explodes. And why it only does this in a certain ratio of the components and does not also pop in a different mixture or even a mixture of similar-looking substances. It is easy to describe how the individual substances are obtained. One can bring the relationship of the substances into formulas. But as long as you don't know anything about the atoms and binding forces, you won't really be able to justify the effect causally. Just because you can deal with something wonderfully and make correct predictions about the consequences that will occur does not mean that you have understood the matter correctly.

The second is radioactive decay. One tries to explain causally how this takes place. Can't we because we don't know all the required parts of the event flow? Or does this causality not exist at all? Is our level of knowledge sufficient to make a decision here? Or can we just continue with a working hypothesis, like in the Middle Ages with black powder? Then, when passing on these ideas, e.g. in the classroom, make it clear that these are only assumptions and do not present them as facts. More on this in Chapter 3.10 on coincidence.

As I explained above, we look at everything with the knowledge we have so far. If one assumes that there is no causal relationship, one will inevitably come to different conclusions when assessing radioactive decay than if one assumes the possibility of a causal relationship.

If you trust your own thoughts and don't want to make any leaps in your own logic, you have to be clear about every single building block of your mental structure. What can be considered actual and what is dependent on further assumptions that do not have to be given. The basics are of particular importance.

These can have different properties. They can be **clear events** and secured within the framework of their experimental process, such as:

- If you let go of a stone, it falls down.

- If you measure the speed of light with light clocks or physically comparable measuring instruments under Einstein's definition of simultaneity, you always get a constant value.

- When examining the spectral light of the stars, the spectral lines are progressively redshifted the farther away their sources are.

But they can also be **unprovable assumptions**, like:

- Gravity makes the stone fall down.

- The speed of light is absolutely constant, so it must also be measured constantly.

- Similar to the Doppler effect, there is a shift in the spectral lines in the universe, so the universe has to expand.

At this point, I only want to address the first point because it leads to the next section. All my life I have learned only in sense impressions, and my thoughts run only in models of these sense impressions. That's why I always saw school as training for language. Today we are being taught language ever earlier (already in kindergarten) and ever more massively. And instead of going into nature and sharpening our senses for observation, we only learn the language of Nintendo, Playstation or computers with an artificially defined framework.

In principle, everything that man has ever put on paper can also be stored and retrieved on the computer. But everything is pressed into a form so that people can exchange this information with each other. This also applies to devices with which people communicate, but which only ever act in a way specified by the designer. Even the most able writer will not be able to properly describe the scent of a rose. Only the rose itself can do that, and we can learn to perceive it through our nose. The flowers can also be fragrant to insects without them understanding human language. All these languages are made only from the knowledge and skills of humans. Not from the infinite diversity of nature. The knowledge that people can acquire and share with each other is inevitably limited by the language they learn through education. Only those who are aware of the barrier created by language education can imagine that there is more behind it.

Nature drops stones, planets orbit suns, and many suns orbit in galaxies. Everything is secured as such events. But why is that? Linguistically shaped people are often satisfied when the child has a name: gravity. I ask: What is gravity anyway, what is behind this combination of letters? What is behind the mathematical description of the general theory of relativity?

### 1.3 The linguistic or the mathematical reality

Linguistic terms can easily build bridges so that we don't stumble over the spider leg problem described above in our minds, but simply walk over them. Man in his everyday experience does not think beyond the linguistic concept of gravity. But even the formulaic description of gravity in mathematical language is not enough for me, just as I would not be satisfied in the Middle Ages with just the description of which components you have to mix together to get a substance that makes a bang. I think we know too little about gravity itself. Only when we have clearer ideas about their character in the reality of the universe around us can we improve the formulas that describe them.

In order not to jump logical gorges in our own thoughts, we must always be on the lookout for the spider legs. In areas that can be controlled by human senses, where we can see or touch the spider legs, these problems hardly exist any more. Someone has already discovered the mistakes.

In atomic physics and astronomy, however, the experiments are more **black box events**. On one side we throw something in, and somewhere else something else comes out. We cannot directly observe what happens in between. As a result, when we change an experimental procedure, we cannot see whether, by cutting off the spider legs, we have also changed conditions other than those we had imagined.

How to approach this problem? Here we come to the second reality, the linguistic or mathematical reality. Since we cannot observe the process itself in the case of black box events, but only its effect, we can only establish a formula-based relation. For example, in the case of antihypertensive drugs, where we see a stronger effect when the dose is increased, but we cannot see from this relationship why the drug works.

In physics, the conditions are more uniform, so the measurement results are more uniform, but the knowledge about the process itself is no better. If these conditions produce different results under what appear to be the same conditions, we are left with only statistical statements with probability calculations. These can fully correspond to the mathematical laws and offer a mathematically self-contained logic. However, this only corresponds to what I call a *linguistic or mathematical reality.*

Understandably, these formulas do not reflect the different prerequisites that are hidden from us in the black box, but which nevertheless exist in the reality of the universe that surrounds us.

I once read a bizarre story of how a man, increasingly free of gravity, finally clung to the top of a pine tree, lost his footing, and then disappeared into the sky. The process is described in a captivating way, but it is not real, it is only real in language. During my studies, in the course of medical statistics, the example was often given that probably at the beginning of this century the development of the number of storks coincided with the number of births. The statistics should show that the children are brought by the stork. But that's just a mathematical truth. Mathematics cannot help to answer the question of whether the children are brought by the stork. It is much more important to answer whether this mathematical connection actually exists through natural conditions.

An example that perhaps theoretical physicists and specialists can also understand why mathematics is not nature itself. For observers who are in a box, according to the principle of equivalence, there is no way to make a measurement inside the box with which they can determine whether they are standing on earth or moving through space with constant acceleration. The measurement results and measurement methods are the same, above all the math is the same. One has to make observations independent of this mathematics, like looking out the window. Only then do you see a difference. One observer is still sitting on his balcony after 100 years and sees the sun moving across the sky during the day and a starry sky that is constantly distributed evenly in all directions at night. The other sees a completely distorted universe pass by at almost the speed of light after only one year and energetic problems arise to maintain the acceleration. You have to look at these as completely different physical conditions. However, this cannot be seen from mathematics alone. One can generalize and say that in the case of black-box events, it is generally not possible to recognize from the mathematics alone which conditions actually exist in the reality surrounding us.

In the case of the back-box events of physics, as I already called them above, it is also much more decisive whether the assumed connection exists at all than their statistical or formula-based description. Even if you cut off the legs of 100 spiders, you won't get any closer to the problem itself. One becomes even more irritated when, with a certain error rate, the spiders' legs are not completely cut off, but one does not recognize the error and some spiders then jump in a completely different way. The two test phases differ not only in the change desired by the observer. The description of the experiment can therefore not contain the change that was not perceived by the observer.

In terms of physics, it is always assumed that the experimental setup is the same at the beginning, apart from the desired change. But in an area that cannot be observed, different starting conditions can still exist. The other initial conditions are caused by something that plays a role but cannot yet be observed by us and therefore cannot be included in the considerations.

This can also affect a general process in which one finds a difference between the measured value and the calculated value, such as the Pioneer anomaly, the fly-by anomaly, or galaxies rotating too quickly. A meticulous examination of the individual parts of the sequence of events has led to the solution of the Pioneer effect. The formulas were correct, but the asymmetrical radiation of the energy was not correctly estimated. For the other problems I would like to present an alternative gravitational model later as a solution. Here, too, the formulas do not need to be changed, only the basis of the measured values on which they are based should be evaluated differently.

This second level of reality can be a mathematically appealing, logical and self-contained world. Assumptions must be made for the basis of these mathematical worlds. Assumptions like: there is no absolute space, there is no absolute time, there are inertial systems, and the speed of light is absolutely constant. In the next chapter I will break down the statements about space and time in more detail.

Since the assumptions do not have to be real components of the universe that surrounds us, their mathematical conclusion does not have to be a component of the universe that surrounds us either. But even if the assumptions are correct, the mathematical description has no more to do with the universe than an image description has to do with an image itself.

A painting itself is complete in its existence, in its temporal and spatial origin, in its causal sequence of the production of its materials and in the composition by the painter himself. However, the description of the picture is a linguistic work of the author with its own existence and not that of the image itself. The person describing the image can only include in the verbal description what he himself has already observed and only in the quality of observation and description that he is capable of himself.

Imagine if a disciple of a Great Master made an exact copy of one of his works. The student could have worked on the master's work as well as the master on the copy. The existence of these images is known from the descriptions of the images. But these were long lost and are now being rediscovered. Since both were made in the same studio, they are made of the same material. Their age of a hundred differs by only a few days. Under these conditions, a person describing a picture will no longer be able to determine which picture was made by the master and which by the student, even with the most precise examinations. Therefore, in its linguistic reality, it will be indefinite which comes from the master and which from the student. In natural reality, however, I believe it is very clearly determined which image or parts of an image come from the master and which from the student.

It is no different with the mathematical description of the universe that surrounds us. Only things that have already been observed and only in the quality in which the observations were made can be included in the description. There are absolute limits for this, which I would like to describe in the following section 1.4, and relative ones, because the observation possibilities are still technically insufficient, e.g. the measuring instruments are not yet accurate enough, or the knowledge is still insufficient, as with the black powder problem in the Middle Ages.

Anything that can be described in this mathematical world is mathematically/linguistically real, but it is not the reality of the universe around us.

I would like to give an example of the mathematical processing of measured values and the conclusions drawn from them. To measure the temperature you can build a tool, a thermometer. Depending on the temperature, the length changes e.g. an alcohol column. Measured values can be obtained in the practical range of about -10 °C to 20 °C. Then you can use a formula to represent the relationship between the column height and the temperature. That gives a very reasonable working basis. If I assume that this ratio does not change across all ranges, one can use this formula logically and according to the mathematical laws to calculate the absolute zero point, namely at the zero length of the alcohol column, because it cannot get any shorter. But in doing so, the spider's legs have been cut off again, because this too is only a mathematical reality. You start from certain assumptions and come to a conclusion through the logic of mathematics or the logic of language. In the nature that surrounds us, however, this result is only as real as the assumptions are correct or the mathematically described connection applies at all. Here, however, we have already penetrated through actual observations and test executions into the temperature ranges close to zero and had to realize that such a column does not reach zero length even at absolute zero and the zero point has to be defined completely differently.

This linguistic reality also includes statements such as: "There is no model-independent reality." I think this statement is wrong. The wording "there is no model-independent knowledge" would be better. No matter what theory or model you have, completely independent of any idea, black powder really pops. I would like to use this to lead to the next section, because I believe there is still a model-independent reality that is independent of human thinking. That too is just a belief and cannot be proved by anything, but it cannot be disproved either.

## *1.4 The natural reality*

There is another reality, the so-called outside. In order not to call it "real reality", I would like to call it the natural reality. This is probably the reality Einstein had in mind when he said "God doesn't play dice". This is the reality that contains the conditions by which nature functions. Completely independent of man's existence or man's observation, independent of man's thinking or attempting to describe them in formulas which he then calls laws of nature. Man can only try to approach this reality, but he will never be able to recognize or describe it exactly. He has to accept that, but he shouldn't throw the baby out with the bathwater and say: This reality doesn't exist. An example of this is "Schrödinger's cat". I believe that the state of the cat is always clear in this reality. Only the state of knowledge of the observer about the state of the cat and thus its state in the linguistic-mathematical reality remains undefined until the lid has been opened. This also applies to the physical observations for which it stands as a model, such as the double-slit experiment. In the chapter about wave-particle duality this is described in more detail.

I want to give another example of the limit to this reality: if two lines are at a certain angle to each other, or two particles are moving at a certain angle to each other, the exact value for the angle can fill a thousand pages. And if another particle moves at an angle just slightly different from this, that exact value could encompass more characters than the previous literature written by all mankind. So we are unable to describe this angle exactly in formulas, let alone measure it. God has it easy with the description by simply letting the particles move at this angle. However, the observing person will only be able to approach it up to a rounding that he can handle, and that even at just one angle. That remains the dividing line to this reality for man. If humans cannot exactly describe the angle at which two particles move in relation to one another, then this is not because the particles are not moving exactly, but only because of the limited ability of humans to observe and express themselves.

I would like to close the chapter with a quote from Einstein from a collection of quotes: "*Logic gets you from A to B. Your imagination gets you everywhere.*" (translated by me)

And transform it into a relation to reality: natural reality is a causally self-contained system with clear paths. In the world of language you can go anywhere and describe anything.


# 2. Space and Time

### *2.1 measurement of space*

Suppose there is an absolute space. How can it be determined whether a particle is moving in this space? Someone who could perceive absolute space would then also know whether a particle is moving or at rest in space. How could man perceive this space, what property of this "absolute space" could man perceive? Humans can only perceive something through a measurement. To measure means to compare something within the universe around us. For example, comparing a photon of light arriving on our retina with the state when no photon arrives there. Also the effect in the inner ear when a sound wave acts on our eardrum, compared to the resting state. But how should he measure this absolute space? If a person measures a property, it must be questioned whether it is the property of the "space itself" or the property of a medium located in the room.

Let's take two particles. How can you tell if these particles are moving towards each other? You think it would be easy with rotation. But how can you determine whether one particle is rotating or whether the other particle is flying around it? One will quickly ask about centrifugal force. But even that is not an easy solution. I want to do that later in chap. 5.2 with an example in which two rings rotate parallel to each other. What decides in which ring a centrifugal force is measured, is that just coincidence? Or even: What should compel a centrifugal force to be measured at all?

What about the distance between two particles? It could represent a part of absolute space. But how can humans perceive the distance, i.e. measure it? He could interpret it with standard meters. Now he may find that he can always lay out the same number of standard meters between the two particles. Relative to absolute space, however, the standard meters could shrink, so the distance between the particles would also decrease relative to absolute space. He would therefore have no possibility of perceiving this absolute space.

With the best will in the world, I can't imagine any possibility of perceiving such an absolute space. One can decide to say: what cannot be measured does not exist. It's almost like closing your eyes and saying, "The universe doesn't exist because I can't see it." But when you open your eyes again, it's been there from the big bang to now. And when you couldn't measure X-rays, didn't they exist? That doesn't make any sense to me!

Just because I can imagine it, I think it's possible in principle. It's true that you can't prove it, but you can't disprove it either. In order not to unnecessarily limit the possibilities that can be reached in a logically correct structure, I still consider the existence of an absolute space to be possible, even if humans will probably never be able to measure this space.

What about time? Again, man can neither prove nor disprove that there could be an absolute time. Here, too, he can only compare two variable processes with one another. E.g. two clocks compare with each other or a process to be compared with a clock.

When man describes motion processes in the universe, he finds that it only works in a four-dimensional connection of measured values of space and time at the same time. Measured with light signals (see definition of the meter m) and atomic clocks. The question remains, are space itself and time itself four-dimensionally linked in natural reality just because its measurements are four-dimensional? I would like to go into more detail on this in the following sections. Humans cannot answer this question unequivocally either. The situation is quite different with what humans can observe. Certainly the senses can be used to make certain measurements. But if we want to make exact measurements in space and time, we need exact definitions of meter m and second s.

## *2.2 Measuring the size and distance*

What does measuring mean? E. g. the fundamental quantities space and time? To measure space, distances are determined, e.g. by comparison with a measuring instrument. In order to be able to work properly with it, this measuring instrument must remain constant under all measuring conditions. You can make a standard meter from special alloys that is as constant as possible in size, e.g. shows no measurable difference in length at different temperatures. With greater measurement accuracy, e.g. when comparing with light, you then determine length differences and you look for a new measuring device that does not show any differences within the scope of measuring accuracy. Today's definition is:

Quote from the website of the Physikalisch-Technische Bundesanstalt PTB (no link given, as they are constantly changing):

"The **meter** is the distance travelled by light in a vacuum in a period of (1/299792458) seconds. The meter definition assigns a fixed value to the speed of light[4] c. This fundamental constant can therefore no longer be measured, it is now precisely specified. From this it follows that the unit of length depends on the unit of time second."

There are two main problems here: First, what is meant by "vacuum" or how is it defined? Second, how to measure the time interval? I want to go into that in the next chapter.

How can you bring two similar standard meters (here also in the sense of the above definition) into a "comparable" position under different conditions? And how can I check the constancy for these different conditions? For example, whether the standard meter has the same length horizontally to the earth as vertically to the earth? At first the question may seem nonsensical, but one has to realize that when I assume that the length is constant, it is an assumption. A perhaps more

---

4    Relativistic velocities are often given in multiples of c ~ 300,000 km/s.

descriptive physical example is: If I make comparisons with my standard meter inside the sun, the atomic density must also increase in the standard meter according to the conditions prevailing there. Is "the meter m" also shorter, or is my standard meter now shorter than "1 meter"?

Transferred to vertical to horizontal: If you take two standard meters and place them next to each other in both directions, no difference is found, since the same condition applies to both. What is it like when the horizontal and the vertical standard meter are to be compared? A direct comparison is not possible here. Even with the help of a complex measurement construction, the situation cannot be significantly improved. Firstly, because the same different conditions could also apply to this measurement construction and secondly, because there is always only a limited measurement accuracy. If I measure accurately up to the 15th digit, the effect may not show up until the 23rd digit. Third: Even if I have found such a measuring device that displays different measured values, I am stuck in the relativity dilemma of deciding for which of the measuring devices this direction-dependent (condition-dependent) size difference exists: The "standard meter" - it would then not be in both directions 1 m long - or the newly found construction. Both could also shorten, only one more than the other. Whereby then, from the point of view of the increasingly shortened scale, the other would have lengthened.

Apart from such considerations, **measuring a size** by comparing it with a 'yardstick' initially appears unproblematic. But if a carpenter, when measuring, first reads one end of the folding rule and then slips until he reads the other end, he will never be able to make any useful measurements. It must therefore be required that there is no change between reading one end of the folding rule and the other. Nothing changes in a static structure, so this condition is met. If you check both ends several times when reading the folding rule, you always get the same readings.

But what about motions? Nothing is static in the universe around us, even the continental plates move against each other. How can I determine the length of a body moving along my ruler? Here I can say: By comparing the beginning and the end "simultaneously" with my dipstick. However, the length measurement of moving bodies is dependent on how far a spatial simultaneity of distant locations can be determined at all. The measured value for m is therefore two-dimensional as a function of time, i.e. the space that can be measured by the human observer is four-dimensional as a function of time. This is also expressed in the above definition of the meter m, which is therefore dependent on time. Whether the "space itself" in natural reality, which is independent of the human observer, is also dependent on time cannot be judged by the human observer. Of course, one can think about this space in a philosophical sense. The results of these thoughts, however, always remain limited to the reality of the I and the reality of language. It is not possible to establish a relationship with "space itself" through measurement, so it is useless to try to capture this space. But then you have to limit your conclusions and statements that you draw from the four-dimensional measured values to this measurable reality.

In principle, it can be said that the human observer can only make "comparisons" to determine sizes and distances and cannot determine the "absolute" size. In order to be able to make comparisons, a measurement instruction must be defined. Conclusions from these measured values are then only valid within the framework of this definition. It is important to be clear about the principle of the question, since this plays a crucial role in observation experiments in the field of atomic physics and astronomy.

## 2.3 Measuring Time

The measurement of time is also only a comparison, here of two changes in space. On the one hand the "clock", which is supposed to represent "constant" units of time, and on the other hand the process to be observed. Time is a little different than distance. The "measurable time" only elapses when something changes, e.g. the "clock is ticking". If the clock stops ticking, no "measurable time" is passing either. If I can observe this, I am changing, so time is still running, only the clock has stopped. If there were no changes in this universe, no "measurable time" would pass, because in the truest sense of the word all "clocks" would stand still. But this must apply to everything in this universe, because otherwise the "time" for the universe continues to run, and the clocks have only stopped in certain regions, but an observer from these areas can later determine when it changes again and so that his "ability to observe" continues, that something else in the universe has changed and that "time" must have passed with it, even if he has not changed during this "time" and his clocks have stopped.

There can therefore be an absolute time when the processes in space can pass faster or slower, i.e. an event sometimes requires more or less "time itself". Man can only measure a change in time if this change does not apply equally to all processes in this universe. It makes sense to initially regard time as constant, but the more recent considerations of the Big Bang theory also call this into question, at least for the first few moments.

Just as simultaneity plays an important role in determining distance, simultaneity must also be asked about time. Here the question has to be answered whether a certain event is simultaneous for all observers or whether there are different times for an event. As an example: Did an event happen at the same time for a direct observer (sitting in a café and watching the square in front of which the event is taking place and looking at the clock) as for someone sitting in the same chair three days later read about it in the newspaper with the time stamp when the event is said to have taken place?

One step further: Did it happen for both of them at the same time, even if the wrong time was given due to a misprint in the newspaper? The newspaper as a means of information perhaps makes the question clearer as to whether the means of information = means of observation = measuring instrument may have an influence on when the event happened for an observer. What is asked here is the temporal linearity of a place and thus also the causality. This question should be considered uniformly as far as one accepts the world lines of observers or of light in general.

World lines[5], no matter what physical appearance, are linear and causal. That means: "For every place on the path of this phenomenon that lies between two places on its path, the corresponding time of the clock is also between the times that the clock has indicated at the other places. This is also one of my basic assumptions, which I am not willing to deviate from. I assume that the universe around us is structured like this in natural reality, even if people cannot always measure it that way.

It is more difficult to answer the question of how it is with the simultaneity of events that take place in different places? You can send out a time signal, have it reflected at another location and receive it again. According to the causality of the world line of the time signal, one can say that the reflection event happened after it was sent out and before it was received again. You can look for a faster and faster time signal, but in principle it is not possible to narrow it down more precisely than within the time loop of the fastest signal. With the

---

5    [22] S.79f, [24] S.63

additional assumption that the speed of light is absolutely constant, one can then conclude that the time signal sent with light takes the same time for the way there as for the way back. The reflection event would then have happened exactly halfway through the entire time loop, according to **Einstein's definition of simultaneity**[6]. [9][10][17][24]

In natural reality, space could also exist without the influence of "time itself". Analogies are not evidence, but they can help explain what is meant. Imagine a railway system where the electrical voltage is slowly increased. Then the trains run faster and the lamps shine brighter. However, the observers involved in the system would not notice anything because their clocks also ran faster. The train would still need the same reading of time for its round, and the lamps would not be brighter because the same number of photons per unit time would still be measured. This example should not be overused, because not all processes change equally when the voltage is increased. Let's imagine that the processes in the universe around us would all proceed evenly faster. We human observers would not notice anything of this, because no difference would be measured when comparing any processes. However, this only applies if it applies equally to all physical processes. If the speed of certain physical processes has changed differently than others over billions of years, an effect would be noticeable. For example, the speed of the processes in atoms could accelerate and thus the frequency of the frequency transitions could increase. Because of the laws of conservation of energy, however, this should not lead to changes in the photons that are already on their way. See also Chapter 3.6 Conclusion space is expanding.

Man in his limitations can only observe space depending on his ability to observe and think about it, i.e. dependent on time. He can only perceive this time through spatial changes. Be it a pendulum swinging or a caesium atom in an atomic clock. So in his I-reality he can only observe a space-time continuum. Its representation in the linguistic, mathematical reality can only take place in space-time dependent, i.e. four-dimensional formulas.

---

6   Several physically equivalent methods for determining simultaneity are presented. Einstein[9] p.31: "Two events taking place in the points A and B of the system K are simultaneous if they can be seen simultaneously in the centre point M of the line $\overline{AB}$. Time is then defined by the quintessence of information of the same nature, clocks which are at rest relative to K and which at the same time have the same "hand position"."; but also "... clocks ... arranged according to the following scheme. If a ray of light is sent from one of these clocks $U_m$, when this clock shows $t_m$, through empty space to another clock $U_n$, which is at a distance $r_{mn}$ from the first, the clock $U_n$ should have the time $t_n = t_m + r_{mn}/c$ when the ray of light arrives." [17]Marder represents the method I described.

## 2.4 What is a second?

The more complex our social structure became, the more precise the ways of stating time had to be. In the beginning, a calendar was enough, but with increasing industrialization, precise times of day were also needed. So the day was divided into 24 hours, the hour into 60 minutes, and the minute into 60 seconds. So the second was a specific part of the day. Apart from general problems of the time, what is a second?

From astronomical records of the last 2700 years it could be determined that the day must have become longer and longer. The second defined in this way would also have changed during this time. On average only 17 μs per day, but in total it is about 6 hours, so it can be recognized from the time of day with rough descriptions of the sky observations.

In order to measure such short times, you need very precise clocks. With the atomic clock we have found a very precise measuring instrument. It can now be determined that there are much coarser fluctuations in the earth's rotation speed. On the one hand seasonal fluctuations, but also others, the cause of which is not exactly known. This means that the old definition of seconds has become too imprecise. A new definition has been established. Also taken from the PTB homepage:

"The **second** is 9,192,631,770 times the period of the radiation corresponding to the transition between the two hyperfine levels of the ground state of atoms of the nuclide $^{133}$Cs."

To put it more simply: 1 second s is a certain number of oscillations of the caesium-133 atom. But that's not enough in terms of precision. A clock runs faster on the mountain than in the valley. In addition to this effect, the atomic clocks in the satellites of the satellite navigation systems run slower according to their motion, corresponding to the Lorentz transformations. What physical condition of the caesium-133 atom does the definition now refer to?

1. Is the second shorter on the mountain, or does the clock go faster on the mountain? By definition, the second on the mountain is shorter, but does that make any practical sense? If you implement the definition equally at all positions on earth, the clocks all go differently.

2. A clock resting on the preferred rest system of the Global Positioning System would also run faster than one moving with the earth's rotation. The effect also depends on the latitude at which the

watch is located. The effect must also be taken into account for the satellites moved in the satellite navigation.

Which is the right second now? The Universal Time Coordinated UTC, the world time so to speak, is formed from a clock ensemble in which the clocks are all provided with correction values. Thus the clocks show a different number of oscillations of caesium 133 as a second than the number corresponding to the definition. So most, or all but maybe one, go wrong compared to the definition.. And what if the environmental conditions for the reference clock change?

Even with that, we didn't get the problem of "time itself" under control. But there is another fundamental problem:

Consider two precision pendulum clocks synchronized at sea level. We leave one at sea level and transport the other at 3000 m. If we compare the clocks, we can see that the clock on the mountain runs slower. Does "time itself" pass more slowly here, or has my watch only become slower because of the different physical conditions? If you do the same with atomic clocks, you will find that the atomic clock runs faster on the mountain. With this clock, too, one has to ask whether the "time itself" goes faster, or whether my measuring device only makes the clock go faster because of the other physical conditions. Pendulum clock and atomic clock behave in opposite ways. It turns out that the behaviour of the atomic clocks, in the formula structure of the Lorentz transformations, which was found to be correct, can be better combined with the other astronomical measured values. They also correspond in behaviour to the light clocks. Light clocks and atomic clocks[7] that are comparable to them indicate the time within the geometric framework of the Lorentz transformations. But are they showing the **correct time**?

"Time itself" could pass the same for all observers, only the clocks go at different speeds depending on their type and the environmental conditions.

How is it practically handled with the clocks in the world and implemented in particular in the satellite navigation? Here, all clocks are recalculated to a uniform time, Universal Time Coordinated[8] UTC. Here, a uniform time is used for calculation and the clocks are treated as if the speed of the caesium-133 oscillation were different, depending

---

7    In principle, they are also based on light

8    [22] p.18: *Currently the TAI calculation is based on the averaging of the local time scales of about 150 globally distributed cesium atomic clocks.* The principle has not changed even today. Further information on UTC can be found on the homepage of the Physikalisch-Technische Bundesanstalt PTB www.ptb.de.

on where the clocks are located and at what speed they are moving. It has been found that satellite navigation also has a preferred rest frame to which the earth rotates. One does not bother with this preferred rest system, because it is a rotating system, not an inertial system, and it should also only be a local effect.

It does mean, however, that a light signal sent in a westerly direction between two observers resting on the earth takes less time than a light signal sent back between them in an easterly direction.

One could also position mirrors along the earth's orbit around the sun and send a light signal in the direction of the earth's motion and simultaneously in the opposite direction, which would correspond to a large Sagnac experiment, thus a rotating system. Here, too, there must be a preferred resting system, because a light signal sent in both directions of the orbit can only reach one observer at the same time. This is not possible with all the observers who are moved to do so.

In principle, the local group Earth with its navigation satellites corresponds to a huge Michelson-Morley experiment (MME). The geometric conditions in the universe surrounding us undoubtedly correspond to the Lorentz transformations. This means that the Sagnac effect that occurs when orbiting the sun also has no measurable effect within satellite navigation. More about this in chapter 5.4 about the method of measuring the motion relative to the gravitational field. Earth's rotation is detected, but linear motion is not.

I would like to close this chapter with a quote from Einstein. He was able to express very complicated things very simply. With the help of this quote, I would like to make it clear where the differences lie for space-time in natural reality and in linguistic reality. Einstein was able to explain to a reporter what the general theory of relativity achieves in a single sentence: "*It used to be believed that if all things disappear from the world, space and time will remain; but according to the theory of relativity, time and space disappear with things* " [19]. I would like to put that into perspective and say: If all things disappear from the world, then measurable time and space will also disappear. However, "space itself" and "time itself" could remain unchanged in natural reality. This does not necessarily contradict Einstein's views. He also said, "*Time is what you can tell on a clock.*" But if there is no longer a clock, then this time no longer exists either. This has nothing to do with "time per se".

## *2.5 Simultaneous Events or One-Place-One-Time-Events*

Einstein called the events when a train passes through the station and observers on the train and on the platform look at their watches and each other's watches simultaneous events. They are the crossing points of the world lines of the observers and the clocks. I would like to formulate them more comprehensively, and the term "simultaneous" bothers me because it also evokes simultaneous events that are spatially distant from each other. In the case of spatially distant events, assigning them to different observers becomes very complicated. I would like to go into this in the next section. Here I want to deal with the events that only happen at one point in space and only at one point in time, so I call them one-place-one-time-events (EOZ).

Two observers A and B meet and look at their watches. For the observers and their clocks, this event only happened at a specific point in space and at a specific point in time in their respective world lines[9]. These are definite events that apply to all observers of this universe. Everything that the observers have experienced before has also happened before for all observers. And what they only experience afterwards also happened afterwards for all observers, regardless of how they assign this event to their own world line.

In principle, this also applies to spatial distribution. It's just harder to portray. Let's take a rubber mat on which we will draw the sun, moon and earth. From Earth, the moon is said to be in the opposite direction to the sun. At this moment two observers meet on earth. Then this spatial distribution is the same for all observers. How they assign the image to their own time and space is another matter.

Let's now drag the moon around the earth with the rubber mat so that it lies between the earth and the sun when viewed from the outside. For this external observer, the spatial arrangement would appear quite different. But for this observer, too, a light signal sent from the moon to the sun would take the same path on the rubber mat past the earth observer to the sun and not the direct path to the sun, which is shorter for him. For him, the light would move in wavy lines in this picture.

Possibly this observer could also move a light signal directly from the moon to the sun, which would appear direct to this observer. That could be equivalent to moving through a wormhole. If we don't want to overturn the laws of conservation of energy, then the earth observer would also be able to send this light signal directly from the moon,

---

9    [22] S.79f, [24] S.63

which enters the wormhole there and then moves directly to the sun without passing through the earth observer again. For the observer on the rubber mat as well as the distant observer there would be the same course of events within the framework of the world lines. There is no other process for any observer of this universe. So also for the signals that are sent through the wormhole. A signal could be sent out where part of it is sent normally over the earth to the sun and another part is sent through the wormhole. If the signal through the wormhole reached the sun earlier than the other signal, then it would do that also for the earth observer. From his point of view, that would be a faster-than-light signal.

## 2.6 Simultaneity of spatially separated events

As I described in Section 2.3, there could be an absolute time. This would then also assign a specific time to each event, and there would be no problem with the simultaneity of spatially distant events. Whether absolute time exists or not, an observer within this universe cannot measure it. The only way we can causally assign spatially distant events is to emit a signal that is reflected at the other event and we receive again. In the world line of the signal, the event of reflection happened causally after the sending and before the receiving and thus also for all observers of this universe. We can't narrow it down any more precisely.

Any attempt to limit this with any auxiliary constructions is always associated with additional assumptions, which do not have to apply as such. For example, a clock transport would be subject to the condition that the frequency of the clock is constant on its way. We know from satellite navigation that there is no such clock. Every clock that is moved on earth moves out of the Universal Time Coordinated Synchronization. Einstein's definition of simultaneity is, as the name suggests, a definition that does not have to be given in this way either, it is only a helpful metrological limitation.

In principle, this also applies if the signals are only transmitted at the speed of sound and no faster signals are permitted for the test procedure and the assessment. In this speed range, however, there are no relativistic effects for the course of the clocks and the meter. Of course, you can narrow that down more with a faster signal, but no further than with the fastest signal available. So now a signal propagating at the speed of light. Within the time-space loop of the fastest signal, the simultaneity remains freely selectable without leading to causal contradictions.

## *2.7 Change of observer position*

It is very difficult to look at a sequence of events that one observer
has observed from the point of view of another observer. Not only the
position was changed very quickly, but also the sequence of events.
Many who think they have presented contradictions of the STR on the
Internet make this mistake. One way to avoid this is to orient yourself
to the one-place-one-time events.

One can represent an event sequence from the point of view of an
observer in a Minkowski diagram. With these, the relative conditions,
as described by the Lorentz transformations, can be clearly represented.
Then all crossings of the world lines of observers and of light signals
represent one-place-one-time events. If one shifts the space and time
axes in such a way that another observer can consider himself to be at
rest, then from his point of view all of these Points must be unchanged.

If other one-place-one-time events then arise, then the sequence of
events is not the same either. When comparing different sequences of
events that are nevertheless presented as the same, any result that you
can think of is possible, but it is not correct. They only exist in
linguistic reality, in which rotten planks in the causal bridge are jumped
over as a result of this equality. But they are not part of natural reality.

## *2.8 Linearity and Causality*

In order to develop thoughts logically, the **basics** must first be
established. These can be very "reasonable" to humans, but there is no
reason to assume that they must therefore be correct. It is important that
they are formulated clearly. Then one should present the logically
developing conclusions in such a way that one can see that they are
only valid as long as the assumptions made about them have not been
refuted. If one finds a reason why one of the bases must be wrong, it is
easier to show which of the conclusions lacks the basis for its logical
development.

One of my basic assumptions is the linearity of the world lines. I
think there is a material or physical causality in natural reality. These
are the world lines of individual identifiable structures. These can be
individual atoms or molecules, but also observers, clocks and the
propagation of waves in a medium. The world lines are linear. This
means that for a moving body, for two of its spatial locations on the
world line with the associated times, a location can always be specified

that lies in between, and the associated time then also lies between the other times. So there are no leaps in space or time.

In the following, I would like to limit myself to the transmission of information/time signals that can be clearly identified spatially and temporally by observers on their linear propagation path.

In Chapter 2.5 I already mentioned the wormholes, which I consider to be only mathematical consequences that have not yet been observed. Such wormholes would allow a world line for information signals to causally propagate faster than the speed of light. This could limit the spatial simultaneity more than with light. Thus, a setting of the clocks, corresponding to a spatial simultaneity according to Einstein's definition of simultaneity, would not be causally possible for all observers. Or causally, not all observers could measure the speed of light equally in all directions.

Thus, such wormholes or any physical causal phenomenon allowing information signals faster than the speed of light are not compatible with an actual constancy of the speed of light to the observer, but only with the constant measured value of the speed of light with light clocks under Einstein's definition of simultaneity. In chapters 4.4 and 5.2, I have shown what it looks like with rotation, in which one-place-one-time events can be achieved.

# 3. Fundamental problems of observation

## 3.1 *Influence of education on the interpretation of an observation*

Many respond to what I have described above as the reality of the ego. Let's contrast it with the outside world and make it clear that we only observe the outside world through the lens of our upbringing and education and that this is essentially shaped by **everyday experience**. But what does this everyday experience look like? In everyday life you have a task to do, you want to be successful in it, and that is quickest if you don't question every move and constantly think about whether things couldn't be done differently.

Some who have never learned to typewrite but must use a keyboard develop tremendous speed with the two-finger search system. Certainly the ten-finger blind system would be the better one. But if you try to learn it now, your performance drops drastically, so you stick with the old method.

To solve intellectual tasks, you learn a different kind of action building blocks, with which you can quickly reach your goal. Here, too, the constant questioning of these building blocks would cause uncertainty and slow down the pace of work. The more successfully you can deal with such building blocks, the more successful you can be with your argument. However, the greater the risk of skipping theoretically over the rotten boards mentioned above..

If you want to cross a gorge over a suspension bridge in natural reality and assume that all boards are stable, is taught better at the first mistake and fall into the abyss. So he will check every board, which is why he takes a very long time to cross the bridge. The second time, however, he will trust that the boards will hold up and will no longer check every board.

In the linguistic reality of the sciences, however, this is much more difficult. Here the bridges are built only from theoretical boards. These can also be checked individually by questioning them. But only in the overall context do they form the bridge. They must not only be checked alone, which is done mathematically on a regular basis. They also need to be evaluated in the context of the whole. As an example, the twin paradox (see Chapter 4.6). Viewed in isolation, both twins are mathematically completely equal. But it doesn't matter in which direction the space-faring twin moves. Due to its inertial system, not only does the other twin move at its cruising speed, but an entire universe, which is perceived as distorted as a result. If you look at the whole thing only mathematically limited to the twins, you don't notice it at all.

If you make a mistake when evaluating one of the boards, as in the case of the spider legs problem, you have a rotten board in the bridge. When theoretically crossing this bridge, however, one will not fall down as in natural reality, but theoretically reach the other side without any problems.

Here we are with Einstein's statement, which is just as valid today as it was 100 years ago (should be from the obituary of Ernst Mach in 1916):

*"Notions which have proved useful in the ordering of things easily acquire such authority over us that we accept them as immutable facts. They are stamped as necessities of thought. The path of scientific progress is often spoiled for a long time by such errors made*

*impassable. It is no idle gimmick, therefore, when we are trained to analyse the current terms. It breaks their all-too-great authority."* (translated by me)

In principle, all non-verifiable assumptions and basic ideas can become such dogmas. Be it that the speed of light is absolutely constant, the general red shift is due to an expansion of the universe, or whether the negative outcome of the Michelson-Morley experiment (MME) leads to the conclusion that there is no ether. Questioning such dogmas must not degenerate into driving on rails, which already leads to the well-known goal with the question.

It is forgotten that **Einstein's postulates**[10]: "the measurement of the speed of light results in the value $c$[11] in every direction" and "no physical measurement (also non-mechanical) can determine a fundamental difference between the two inertial systems K and K' "[12], a substitute for assuming an ether[13] is. Einstein logically derived the Lorentz transformations from his postulates. Lorentz himself derived the transformations named after him from the negative output of the MME and the assumption of an ether. There is equality of ideas here.

The ether as the basis of the thought structure was rejected because it was supposed to have properties that could not be found. But how far are we today with the properties of light and gravity. There is still a wave-particle duality. And what is the gravitational field? Waves are supposed to spread here. Is that supposed to be a change in space-time. But then the space-time must have some kind of structure, to which an observer can then also move. (See also chapter 3.8 about waves and particles) Relative to this medium, light could then also propagate with constant speed. With the same measuring methods and the same formulas, the same ratios for the measured values would result here. In principle, the initial conditions are comparable to those of the box that stands on the ground or is accelerated uniformly through free space. We must consult other observations to decide which is the correct solution.

By no means do I want to fundamentally question the **principle of relativity**. From Wikipedia it says: "*The principle of relativity states that the laws of nature have the same form for all **observers**. Simple considerations show that for this reason it is impossible to determine a preferred or absolute state of motion of any observer or object. This*

---

10  Also called axioms of the special theory of relativity [4] p. 891

11  Constancy of the speed of light [9] S.30, [10] S.11, [11] S.13, [17] S.19, [24] S 13

12  Principle of relativity [11] S. 4, [10] S. 8, [17] S. 11, [24] S. 12

13  A medium to which the principle of motion could be applied

*means that only the motions of bodies **relative** to other bodies can be determined, but not the motions of bodies relative to a **preferred reference system**.*"[14]

I think it is correct insofar as humans cannot determine any motion to an absohlute reference system and thus an absolute state of rest. That is why it sould be written "to an **absolute** reference system".

Without a frame of reference, it is not possible to take measurements that are spatially or temporally distant from one another. Depending on what you want to measure, however, there may be a preferred reference system. E.g. a state of motion in which the size assumed to be the motion dipole is subtracted from the CMB. This would be a preferred idle state over the CMB, but not an absolute idle state. But this is about the field to which the gravitational waves are moving, i.e. the gravitational field.

The speed of sound is determined by the medium through which it is transported. This medium can therefore also be described as the preferred reference system. For example, if air or water is moving against the earth, the satellite navigation will measure different velocities for the outward and return paths of a sound signal, even if the speed of sound is the same in both directions to the medium. This preferred reference system (sound medium) is therefore only valid for the observed phenomenon of sound motion. The addition theorem does not apply to sound either. The sound only moves at the same speed in all directions to an observer who is stationary in the medium. And the speed of sound is independent of the motion of the source or receiver.

As an example, imagine measuring the speed of sound in the pressurized cabin of a Concorde supersonic aircraft. We send a sound signal from the stern to the cockpit. Then the plane takes off. The measurements are taken again and there is no difference in the readings, even though the sound signal is travelling at supersonic speeds relative to the earth. No new laws of nature are required here, and the principle of relativity is not invalid either.

To measure, we need measuring instruments. If we want to compare readings that were taken at different locations or times, we also need measuring instruments that measure in a comparable way at the different locations. If we want to measure motion processes, then we have to measure space and time at the same time. What about the clocks alone? There are many clocks in the satellite navigation systems, such as the Global Positioning System (GPS). And all the clocks tick

---

14  There are good explanations for the marked terms in Wikipedia

differently when they display the defined second, depending on their position in the gravitational field and their motion, e.g. in the satellite. They are all trimmed to the standard time Universal Time Coordinated (UTC) or International Atomic Time (TAI) via correction values.. This system can be wonderfully navigated as a reference system. Here I see a preferred reference system to which the motions of bodies can be specified exactly.

The microwave background radiation is a frame of reference to which motion can be measured. But that doesn't really help us. We do not know how the microwave background radiation is related to the gravitational field. It could contain a drift and move against the gravitational field. This would measure motion to the microwave background and not motion to the gravitational field. However, the gravitational field could be the decisive frame of reference for the motion of light photons.

## 3.2 Influence of the observer on the experiment and its result

There are many observations in modern physics that cannot easily be fitted into the existing structure of thought. Quantum physics brings astonishing things to light in attempted solutions. In these, much is hidden in the vagueness of things. A vagueness that made Einstein and Schrödinger uneasy throughout their lives.

Quantum physicists say that everyday experience is unsuitable for judging processes, large and small. That is why we could not infer the conditions of the micro and macrocosm from everyday experiences. In our everyday life, everything outside of humans appears to be determined. Man just has to observe things and experience how they are. In the microcosm, however, things are only determined by the experimenter's measurement; without his measurement they would be indeterminate.

What are the experiences in the everyday realm accessible to the senses? Imagine you want to observe a family of lions, e.g. sunning in front of the lion's den and how the young play. I assume that it is a clearly defined process that takes place in this unambiguous and never indeterminate form without human observation. Still, as I walk among the lions with my whirring camera, I won't observe any of this, but be glad if I make it out alive. But that has nothing to do with the fact that the process would have been indeterminate without this observation, rather I changed it through my observation. If I want to observe how something happens in natural reality, I must not change it through my observation.

I just put a basket with a whirring camera in a place where I assume the lions will later bask. So it can still happen that the lion attacks the unknown, whirring basket and shows aggressive behaviour, which it would not do without my observation instrument. An experience made by a wildlife film-maker.

You have to keep the influence of the observation as small as possible. And you have to be very critical at all times as to whether you have really grasped all the possibilities of influencing your observation on the observed process. Only then can one judge the extent of the change in the observation result through the observation.

Let's take another example from inanimate natural reality: we are trying to judge the motion of a falling stone. To do this, we install two light barriers one above the other. We drop the stone above the upper light barrier so that it passes both light barriers. You can change the distance between the light barriers or use several light barriers and thus obtain a wide variety of measurements of the falling motion of the stone. However, the point here is not to assess the falling motion of the stone itself, but to the principle of measurement.

When a body is irradiated with light, it undoubtedly changes its state of motion. If it is an atom alone, this effect is very large. If it is a large stone, the light barrier only changes the motion of the stone below the measuring accuracy. As a result, the process does not seem to be influenced by our observation. However, we must not be mistaken that our measurement does not change the process. Except for cases like the observation of the lion family, in everyday life, by observing a process, it is usually changed only below the measurement accuracy, but it is not different in principle from observing the motion of an atom. We can take this with us from everyday life: No matter how much effort we put in, a process is always changed simply by observing it. In everyday life, however, we can relatively easily keep this change below the measurement accuracy, which means that it no longer affects the description in linguistic reality.

What would it look like if we didn't have such options as a light barrier and could only determine the motion of the stone by the fact that it moved another stone when passing the measuring point? In the event of a collision, it would change direction and not even reach the lower barrier. Of course, one could now philosophize about it and invent a relation that says that one can only measure a stone either at the top or only at the bottom, but not at both. In everyday life we can open our eyes again and see that the stone has taken a different path through our measurement. You can also get closer to the effect with your eyes closed by paving the whole floor with measuring points and then measuring the arrival of the stone. But you only do that if you are concerned about the whereabouts of the stone and are not satisfied with the fact that it might have dissolved during the first measurement. A somewhat provocative interpretation, but I think such macroscopic examples could help to remain critical in the interpretation of light phenomena.

If you take a lamp that emits single photons, then there is a huge bouquet of questions

How can it be measured when the photon takes off?

How can I measure where it flies along?

How can it be ensured that a photon measured at the first station is still the same as the one measured at the second station?

And I ask: Can one consider the path and the timing of the motion to be undetermined just because there is no way to date to observe the motion of a photon without influencing the process in a very massive way?

For this I would like to change the experiment from the area accessible to the senses in order to clarify the problem somewhat. Each stone can be marked with colours or symbols, and even if a whole sackful of stones is sent on a journey at once, a high-speed camera can follow the specific, unique path of each stone, which would be completely impossible with a laser pulse for each photon.

Now let's drop a drop of water instead of the stone. The stones should be at the measuring stations and not light barriers. We measure the effect that the water drop leaves on the stone as a measuring station. Because the stone is still dry, the drop gets stuck on it. No matter how many measuring stations are set up on the floor, the drop is no longer measured because it has actually dissolved, so to speak. The second drop mixes with the first on the stone and drips off it again. This can be measured at the lower station, but even if it contains the same number of molecules, it is no longer the same because when it was mixed with the molecules of the first drop on the first measuring stone, the composition was different. The drop that arrives at the second measuring station is therefore equivalent to the drop that was sent, but not the same. It becomes clear when the first drop contains blue dye, the second yellow dye and a green drop arrives at the bottom on the second measurement.

If a green light photon is sent and a red one is measured, you know it's not the same, but changed. If a red photon is sent and a red one is also measured, how can you be sure that it is the same photon and not just an equal photon? It becomes even more difficult when an astronaut sends off a green photon with his photon lamp and the observer who stayed behind measures a red photon. Is it the same photon? Also, if we assume it's the same photon, is it equivalent? Here we are still far away from relativistic problems. The same question also arises in everyday life, even if we don't really ask ourselves the question. A boy throws a stone at a window. That just makes a sound. Now he rides his motorbike towards the pane and throws the stone again using the same technique, breaking the pane. Was that the same stone now? The answer to the question is clearly yes, but is the stone equivalent?

Conclusion: In everyday life, every process is influenced solely by the fact that we observe it. Here we can usually bring the influence in a subliminal size for the description of the observation result without any problems. In principle, the same applies to the microcosm, except that here the influence of observation always changes the observation result by a relevant magnitude given our current possibilities.

### *3.3 The actual observation*

From the "reality of the I" a process of natural reality is observed. An apple falls from the tree. All apples do that, none of the apples float to the sky. The greater the height from which he falls, the harder the impact. When the apple is thrown into the air, it moves in a specific path called a parabola. In the sky, the planets and comets move in specific orbits around the sun. The better the means of observation and measurement, the more precise the measurement results. Several such actual observations shall be listed here as examples.

1) The **Michelson-Morley experiment** is also such an actual observation. The test setup is clearly defined. And no effect was found with any of the experimental setups comparable to the Michelson-Morley experiment. One can assume that he is not in a position to measure something from the outset, because if you move the mirror on one of the measuring arms, you can see a clear change in the interference fringes. You can also make one arm very short, which does not change the negative outcome of the attempt. It corresponds to the

Kennedy-Thorndike experiment[17]. You can also omit the arm completely, which corresponds to a light clock.

If the relative conditions in the universe surrounding us correspond to the Lorentz transformations, then both experiments must also turn out negative for acceleration in a rocket.

However, the experimental setup can be further reduced to a **one-way Michelson-Morley experiment**. Here, no mirror is mounted at the end of the arm, but a different laser. The signals from both lasers are brought to interference at both ends. With the precision of today's clocks, you can also build clocks at the ends of the arm that send out time signals and then compare them to the clock at the other end. These two times can then be sent back to the other clock. This experimental setup also remains negative when rotated slowly.

Acceleration in the direction of the axis of the experiment has no effect on the return of its own signals, which were reflected by the other clock. The time difference remains constant. But the clock at the front in the direction of acceleration determines that the time signals are arriving later and later, and the clock at the back that the signals are arriving earlier and earlier. This is also noticeable in a change in the time difference that occurs between the reflected time signal and the time value from the clock at the point of reflection. This follows logically from the Lorentz transformations.

2) The **tone of a sound source** sounds lower as it gets farther and higher as it gets closer. But also if the listener moves away from the sound source, he hears the sound lower and if he moves towards it, the sound gets higher. The same applies if a radio or light source moves towards or away from a receiver. Or the other way around, the receiver moves. Depending on the relative speed, the measured frequency of the sound, radio waves or light photons changes. This is also called the **Doppler effect**.

**Light** from an incandescent source can be fanned out on prisms and you get a **spectrum** from infrared to ultraviolet. If the light is first passed through a cold gas, the atoms or molecules in the gas absorb photons of a certain wavelength. This creates black lines in the spectrum. If the gas consists of only one type of atoms or molecules, a line pattern that is very typical of this gas is created. If you mix several of these gases, these line patterns all appear together, but there is no shifting of individual lines. From the line pattern, one can deduce the gases through which the light fell.

On Earth you can send white light through hydrogen gas and study it. The resulting spectral line pattern can also be determined when examining the light of all stars and galaxies. But one can also state that this pattern is **not always at the same wavelength position** when compared with the **spectral line pattern** of hydrogen obtained on Earth. It can be shifted to both the blue and the red light of the spectrum.

Determining distances in astronomy is not easy. But in general it can be said that **the weaker the light from a galaxy** that reaches us, **the further away it is from us** and the more red-shifted the spectral lines are in its light.

All of these observations are **unequivocal observations** and are repeatable in their nature. They can also be made in different ways by changing the experimental setup. In the case of astronomical observations, the experimental setup cannot be changed, but the same result can always be obtained with the same setup of the measuring device. This allows them to be presented as a clear factual observation.

3) If you mix certain substances in a certain ratio, you get **black powder**. If it is compressed in a vessel and ignited, it makes a bang. Another factual observation I'll go into for comparison.

## *3.4 Description of the observation in linguistic reality*

The observations made, more precisely the measured values, can be summarized in formulas, which are then called laws: for the falling of the apple the laws of fall, for the motion of the planets the Newtonian gravitational laws, for the sound phenomena the laws according to C. Doppler, according to him too called the Doppler effect. He had developed these formulas that describe the measured frequencies. In the case of slow sound phenomena that are transported over a medium such as a gas or a liquid, it is found that it makes a difference whether the sound source or the receiver is moving towards the medium. Correspondingly different formulas must be used. With light it's a little different. Let's start with a constant distance between the light source and the receiver. If the light source or the receiver now changes this state by using energy, the effect when the signals arrive at the observer after the change is the same for both. Only if the observer accelerates does the effect occur immediately when the sender accelerates, only when the signals have arrived at the receiver after the sender has accelerated. Here, one formula is all that is needed to describe the change in the frequency readings as they approach or diverge from each other.

With these descriptions or formulas, which are also called laws, predictions can be made for measurements in other situations. In the case of the laws of falling, one finds that they only apply in their form if the falling path is relatively short in relation to the distance to the centre of the mass, which should ideally be point-like. If the path is large, one must go over to the laws of gravitation. But here, too, the masses should ideally be point-like and not rotate. However, planets and especially the sun are not point-shaped and they also rotate.

A more precise measurement revealed a perihelion rotation of Mercury. This is only partially described by Newton's laws of gravitation. So one had to go to Einstein's general theory of relativity, which is able to describe the rest of the effect as well. These formulas also show that the rotation of the earth also causes an effect called the Lense-Thirring effect or Schiff effect[15], which has now also been proven. The question is: has one now recognized the nature of gravity?

A question for comparison: If you use a formula to show how many grams of one substance must be mixed with how many grams of the other substances, has one then recognized the nature and functioning of black powder?

Einstein considered the perihelion rotation of Mercury and the Lense-Thirring effect only to be superimpositions of the gravitational field. I consider them to be dynamic changes in the medium of the gravitational field, to which the concept of motion can be applied. Unlike what Einstein described in his speech on May 5, 1920 at the Reich University in Leiden. About which he might have thought differently at the end of his life. This does not change the formulas with which the measured values can be calculated.

## 3.5 Conclusions from the formulaic representations

Now we come to what is actually not observed at all, but which is concluded from the observed under additional assumptions. Since the assumptions do not have to be correct, the same applies to the conclusions.

The speed of light is always measured as constant with ideal clocks. However, this only applies to the outward and return journey together. The easy way cannot be clearly measured. Under additional assumptions, such as Einstein's definition of simultaneity, one can also measure the simple way, but the measured values are only as correct as the definition. Einstein avoids this and postulates the statement: The speed of light has the constant value c to an observer in all directions.

So far, no measurement has been made that would have determined a difference between K and K'. Einstein also generalizes this statement in his second postulate: There is no physical measurement with which a difference between K and K' can be determined. As is well known, Einstein logically derived the Lorentz transformations from these postulates, with which the measured values of the other system can be calculated.

Lorentz himself assumes that there is an ether to which the Michelson-Morley experiment (MME) is moved. From this assumption and the negative outcome of the MME, he logically developed the Lorentz transformations. Both ideas stand side by side in a logically self-contained manner. The mathematics of the Lorentz transformations alone cannot be used to decide which of the ideas is correct.

Einstein's solution was chosen because the ether should have properties that could not be measured. But where are we today? There is also the wave-particle duality. See also Chapter 3.8.

One believes in the existence of dark matter, which only has the property of completing the formulas, but is otherwise undetectable. Shouldn't there also be the possibility to assume that there is an "ether" that only has the observed properties? E.g. the gravitational field, which in a somewhat modified sense of E. Mach is formed from the fields of the mass particles and which also affects them[15]. This also explains why inertial mass equals heavier mass. As strongly as a mass opposes a change in motion compared to the gravitational field through its inertial mass, it also pulls on it itself, as a heavy mass, like one pulls on the ends of a rope.

It is not just a simple rope, but a three-dimensional rope that fans out (it would be better to speak of a force principle), which is why the tensile force decreases with the square of the distance. In a different way, a heavy mass deforms the gravitational field, which is why it gives an unequal pull to a nearby mass, which is why it moves in the direction of the other mass, or orbits around it.

Because of this force principle, mass particles can only be accelerated up to the speed of light relative to the gravitational field

---

15  [9] p.58: "... E. Mach... According to him, an isolated mass point should not move against space, but against the mean of the other masses in the world without acceleration; ..." (translated by me)

and to which the massless light photons move exactly with the speed of light.

If one admits the principle of motion to the gravitational field and considers it as a medium, solutions that can dispense with dark matter would become possible. Then the twin paradox (see Chapter 4.6) could also be explained logically and clearly. The faster than light speed would then not lead to any causal problems. What is certain is that the absolute constancy of the speed of light is incompatible with a causally acting speed faster than light, no matter on which basis. This would allow a difference between K and K' to be measured. This would refute the postulates on the SRT.

## 3.6 A Special Conclusion: Space is expanding

How observations become conclusions that require further assumptions to be made, but are presented as facts over time.

Conclusions can be drawn from observations together with further assumptions. However, one must always make it clear that the conclusions are only as correct as the further assumptions. If this is forgotten, over time the conclusions can be presented as facts. I would like to illustrate this using the example of space expansion. For this I divide the sections into:

        A) The observation
        B) The formulaic description
        C) The conclusion
        D) The problems

**1) Brightness of a light source (lamp, star, supernova)**

**A)** Observation: First, the observation or discovery itself, which as such consists of a repeatable fact of observation or a sequence of events. Consider a meter used to measure the brightness of a light source and a candle or lamp. In the laboratory, we set up the measuring device and the light source at different distances and **measure** the brightness of each. It is found that the further away the light source is, the darker it appears.

**B)** Formula-based description: Then the measured value of the brightness and the distance can be represented in a **formula-based context**. This can also be checked outside of the laboratory on Earth and even in our solar system. Here we can determine the distances quite well due to the different constellations of the planets. Nowadays even by measuring the runtime of radio signals to space probes and back. Here, too, the developed formula is confirmed again and again.

**C)** Conclusion: One can conclude: If one knows how bright a light source would be on earth, one can determine the distance from its brightness using this formula.

**D)** Problems: Beyond this close range, in which the assertion can also be checked with other physical methods, it becomes difficult to determine the distances. Within the Milky Way or even to other galaxies, it is impossible to determine the exact distances in a geometric way. Also, we cannot receive any transmitted signals after reflection. But it has been found that there are very specific stars that shine with a very specific brightness. If this brightness is always the same for these stars, one can deduce the distance of this star from the measured value of this brightness, also called apparent brightness, and the formula that was developed in the close range. An analogy that is only as correct as the assumption that the ratio does not change over long distances. This assumption could not be checked by mankind using other measurement techniques and therefore the distance to this star must be made clear as an assumption.

I already explained in chapter 1.3 that a formula for determining the temperature with an alcohol column is not valid in the extremes. This also applies to the formula for determining the distance from the brightness, at least in the short range. The brightness increases as the distance from the light source decreases.

$E = 1 / r^2$

with illuminance E in lux (lx), luminous intensity 1 in candela (cd), distance r in meters (m).

The smaller r becomes, the greater the illuminance. If r is only one micrometer large, then according to the formula, the illuminance is already huge and if we approach 0, then the illuminance is infinite. Here we have a mathematical singularity, so to speak. We only need to approach 0 distance from a light source and all our energy problems would be solved.

This is of course nonsense. You just have to try to get as close as a micrometer to a candle flame.

What is nonsensical is that the formula loses its validity as soon as the light source and measuring device come so close that they influence each other. Why shouldn't there also be changes at galactic distances, making the formula no longer valid in these dimensions? Brightness could decrease much more rapidly outside the galaxy and distances could be much closer than previously thought. I don't think that's the case, but until you've verified it with other methods, such distance determinations should only be used as a working basis and not presented as fact. In this case, checking means that you have, for example, travelled at least halfway to another galaxy and measured the brightness of the stars there. To check you need at least a distance determination on a completely different physical measurement technique.

The comparison with another distance determination on a physically identical basis is insufficient, since the same would apply to this. Only when distance determinations are based on physically different measuring methods and these deliver the same values independently of one another is it justified to assume the distances as given. But even then they should not become dogmas.

### 2) spectral lines and their shift

**A)** Observation: When examining starlight, one finds that the further away the light source is from us, the more red-shifted the spectral lines are. An observational fact independent of any theory.

Furthermore, experiments can be carried out in which the light source and the receiver can both be observed spatially more precisely, in particular their distance and their motion. Depending on the motion of the light source or the observer, a red or blue shift of the light can be determined under different test conditions. This observation in itself is also an unequivocal fact.

**B)** Formula-based description: The red and blue shift in the near range, which is repeatedly shown in experiments, is described in a theory that we call the Doppler effect, and this red or blue shift is shown in corresponding formulas.

**C)** Conclusion: We transfer this theory to the observation of the redshift of the spectral lines in distant galaxies and conclude that the galaxies are all moving away from each other. The further away they are from us, the stronger the redshift of the spectral lines, so they move away from us the faster. From this the **conclusion is drawn: space is expanding**.

**D)** Problems: The observer must remain vigilant here, because one could misinterpret the shift in the spectral lines. In itself it seems to be a very simple thing. But it comes with many assumptions that we tacitly take for granted. We assume that:
- that the atoms are constant in their being,

- that time is constant,
- that the speed of light is constant,
- that light photons continue to move unchanged for billions of years as a kind of perpetuum mobile.

But let's imagine **light slowly losing energy on its way**. That would not matter when measuring the speed of vehicles with the laser gun. But on an astronomical scale, if the light has travelled billions of years to reach us, this could also explain the general redshift without the light source and receiver moving away from each other. From my point of view a rather unacceptable solution.

The definition of the meter has undergone several changes over time. From the original metal rod with two notches, it is now the distance light travels in 1/299,792,458 of a second. The meter only remains unchanged if the speed of light and the second are constant.

The **speed of light could be slowing down**. If changing the speed relative to the light source results in a blue and red shift, then slowing down the speed of light should also result in an overall red shift. This could also explain the general redshift without changing space itself. As the speed of light decreases, the number of meters measured by today's meter definition, or the number of light-years, would increase, but the distance would not actually change. It would be like measuring something with a meter and later with an English foot as a measure. Then the numerical values increased, but the thing itself did not change.

The **timing of physical phenomena could also change**. More precisely, the frequency transitions in the atoms could become faster. Then the spectral lines they produce today would be of higher energy than they used to be. From the laws of conservation of energy, the light photons that have already been sent should not go along with this change. This would explain the general redshift by a change in time without any changes in space.

Of course, this means that the atomic clocks also go faster. The light travels a smaller distance in measured units of time of the same length. For me the most likely solution.

Reason: We observe that gases gather into stars, stars into galaxies and galaxies into clusters of galaxies. The space in between is becoming less and less massive. However, a general expansion of space would oppose this, because then the distance between the stars would also increase. What observations can we make in the universe.

The distance between the earth and the moon is measured with laser technology as increasing, at a rate of 3.8 cm/a. If you add this up to a distance of 1 megaparsec, you get a speed of 90 km/s. This corresponds to the space expansion of 74.03 km/s according to the Hubble constant for a distance of 1 megaparsec.

The speeds roughly correspond. But that would lead to a distribution of the mass particles and not to their compression. But if we assume a continuous change in time, the longer the light travels, the greater the effect. Depending on the rate of time change, the measured speed of light would also change. This would mean that the light to the moon and back would take longer and longer. So the moon could actually be approaching the earth and at some point coincide with it. Just like neutron stars or black holes get closer and closer and then crash into each other.

An observation of modern times is the discovery of **voids** (areas of low mass in the universe) and the cosmic web. The **cosmic web** consists of galaxy clusters and superclusters, as well as the filaments in between.

An expansion of space should lead to all mass particles drifting apart, including galaxies. However, if the local gravitational effect leads to the fact that galaxy clusters tend to shrink further and only the free spaces expand, then only such clusters should form. Similar to a jet of water running out of the tap. The faster it falls, the further it has to expand spatially, and the thinner it becomes. But then it also tears open and lumps form (also called drops). There are **no filaments** left.

However, if we have not expanded the space and small cavities have formed due to irregularities in the mass distribution, these could develop into ever larger cavities, similar to the bursting bubbles in a mountain of foam. What remains are the clusters at the points where several bubbles meet or the filaments where only three bubbles meet. The comparison should not be overstated. The drops are formed by the attraction of the water molecules to each other, comparable to gravity. The bubble walls are formed by surface tension. These do not exist with the mass particles in the universe, which is why there are only clusters and filaments, but no bubble walls.

Space is said to expand to 1 megaparsec at a speed of 74 km/s. This means that a cube with an edge length of 1 megaparsec would have had an edge length of only half that 6.7 billion years ago. So 6.7 billion years ago there should have been eight times as many atoms in a volume of the same size as there is now. 10 billion years ago even 64 times the amount, or 11.7 billion years ago even 512 times the amount.

If the space between galaxies is expanding, then 11.7 billion years ago there should have been 512 times the number of galaxies in the same volume. So at that distance, a neighbour of a galaxy would have been much closer to the other. With increasing distance, the galaxies should have an increasing density, or at least the gases and stars should have a much higher density than we find in our surroundings. When surveying the sky, however, this cannot be seen from the reports I have read.

For me, the distribution structure of the galaxies in the universe with these filaments clearly speaks against an expansion of space. Then there must also be another cause for the Hubble constant, maybe the speeding up of time.

## 3.7 Microwave Background Radiation

One might think that the microwave background would be a good frame of reference for motion. Certainly this can be defined as the state of rest when the microwave background radiation looks symmetrical in all directions. By that I don't mean the fine structuring of the microwave background radiation, but rather the blue and red shift that occurs as a result of the Doppler effect. The dipole to the microwave background radiation, which currently gives a relative motion speed of about 370 km/s for the sun. When a spacecraft accelerates from a position where the microwave background radiation looks symmetrical in all directions, there is bound to be a blue shift in the direction of motion and a red shift in the opposite direction.

Also a medium that can be measured, e.g. the microwave background radiation, is of no use to us to determine a motion to space itself. The microwave background radiation could move through "space itself" like a kind of cloud. In the state of motion with symmetrical microwave background radiation in all directions, one would have reached a state of rest to this radiation. In this case, the state of rest in relation to the microwave background radiation would nevertheless be a motion in relation to the "room itself" and not an absolute state of rest.

It would be desirable to be able to measure the motion against the gravitational aether. However, we do not know to what extent the microwave background radiation has anything to do with the gravitational ether at all. In principle, the state of rest to the microwave background radiation could correspond to a state of motion compared to the gravitational aether. Conversely, a state of rest in the gravitational ether could correspond to a motion against the microwave background radiation. The microwave background radiation is therefore not a suitable tool to clarify whether there is also a preferred state of rest to an essence[16] in the case of the uniform translational motion. An essence that firstly causes a preferred resting state for the rotation in satellite navigation, secondly, the clocks in the valley run slower than on the mountain and thirdly, when they change, the measured value for the runtime of radio signals also changes, as with the Cassini effect.

Even though the masses present in the universe and the microwave background radiation may be in different frames of reference, they have something in common for fast motions over 1% of the speed of light. We're accelerating a spaceship, no matter which direction. Then there is a blue shift in the signals from that direction and red shift in the opposite direction, no matter what type of radiation is involved. With increasing speed, the picture of an asymmetrical distribution of the masses and the density of the energy radiation also arises.

Isolated theoretical considerations have a particular flaw. They are not related to the reality around us. If you accelerate the spacecraft up to the speed of light in a gas-free region of the universe, so that there is no friction, the microwave background radiation and all other radiation would still meet the spacecraft as high-energy radiation, and this radiation would hardly be visible to the rear. In this way, the spaceship would only be irradiated with high energy from one side, which would inevitably lead to the spaceship slowing down. This would accelerate the occupants inside the spaceship to the top of the spaceship. If you compare the spaceships in the linguistic-mathematical nowhere, then there is also no microwave background or other radiation. And spaceship and earthling can both consider themselves equals. In the universe that really surrounds us, however, we cannot really accelerate a spaceship without it also changing its state of motion in relation to the universe and the radiation in it.

This raises the question of whether it is even possible in the universe that surrounds us for the systems K and K' to actually be physically completely equivalent to one another in the case of a uniform translational motion? In other words: whether there is such a uniform translational motion and thus inertial systems in the universe that really surrounds us?

---

16  I'm writing essence because I can't tell what it is that determines the preferred state of rest. However, the concept of motion can be applied to this. I call it the gravitational field.

It can't just be motion in which a body doesn't give off energy and to which no energy is added, because that would also apply to a body orbiting the sun, and in orbiting we clearly have a preferred state of rest. In principle, this should also apply to a motion far away from a galaxy, because this could still mean a circular motion around the galaxy.

From everyday scientific experience, we find time and again that all measurements that we make merge seamlessly with the geometry of the Lorentz transformations when we use light clocks or the equivalent atomic clocks under a spatial simultaneity that corresponds to Einstein's definition of simultaneity. There is no doubt about it.

Nothing changes in this geometry either if we pick out the inertial frame that is at rest with respect to the microwave background radiation. From this we could determine the relative velocity to any other inertial system. It would then also have to move against the microwave background radiation at this speed. This inertial system could then also use this radiation to determine how fast it is moving in relation to the other inertial system.

One might think that the microwave background radiation could help solve the twin paradox. Because the clock that has travelled the longer distance and/or faster than the microwave background radiation has always gone slower.

But there we encounter the difficulty of understanding mathematically self-contained systems with our feelings. Unfortunately, this statement also applies to every inertial frame that moves with respect to the microwave background radiation. In general, one has to say that the clock that has travelled the longer distance and/or faster than an inertial system has gone slower. One can also say more simply: Viewed from two clocks, the clock that changed the inertial frame between separation and reunion moved slower.

This will only change if we don't follow absolutely straight paths, but rather curved ones. A circle can also be displayed for each curved motion. With a circular motion we can generate unique events (see Chapter 2.5) that must apply to all observers and are not dependent on a definition of spatial simultaneity that can be different for each observer.

If the other clock B runs slower from the point of view of an observer A, then from the point of view of the other observer B clock A runs faster. As soon as one can represent a circular motion, the conditions are clear. More on this in Chapter 5.2.

## *3.8 Wave-particle duality of light*

When it comes to light, we still talk about wave-particle duality. I think that's only because we always look at physics through the lens of mathematics.

I already explained in the introduction that mathematics is a language that can be used to describe anything you want. You just have to search long enough for the right formula. Figuratively speaking, just because I call a flower a flower doesn't mean that the flower rose is the same as a flower carnation.

According to the principle of equivalence, we cannot tell from measurements and mathematics alone whether we are standing in a box on earth or are being accelerated uniformly through space.

Light can be described using the wave equations and has the same speed as electromagnetic radio waves. In accepted physics, it is therefore concluded that light has wave properties.[6] p.74, [12] p.150, [24] p.6 But is that enough to get closer to the actual character of the light?

Figuratively speaking, when it comes to light, you have to try to look out of the box and examine the properties of light more closely.

Let's start with the description. What is meant by a wave? The classic wave is the water wave. As wave motion, it is the oscillation at the boundary layer of a medium. This also applies to gas, e.g. as a vibration in the atmosphere or solid bodies such as the vibration of a rope.

But there is also the sound wave as a density fluctuation of a medium, regardless of whether it is a gas, liquid or solid.

In all cases we have a medium made up of time-trackable particles, namely atoms. Through this, the vibration mechanism can also be tracked. But you have to do that in different ways in order not to get a one-sided picture.

Let's take the water wave and its interference phenomena. Let's create waves in a water surface, which then spread out in a ring around it. In our everyday experience we only see the ups and downs of the water surface. But that is only one feature, but it spreads very homogeneously over the area. In fact, the water molecules that make up the medium are circling. But also in a very homogeneous form that decreases towards the depth.

Let's go to the double slit experiment. Here we have two sources from which waves emanate. Our everyday experience and unfortunately also our teachers only teach us that with the resulting interference there are places where there are much higher waves/surface fluctuations (maximum) and at other places a minimum, or as it is said becomes: to an extinction. [6] p.84+85, [12] p.165, [24] p.6

But let's not only look at what our teachers teach us, but observe the motion of water molecules. Because, where the teachers say there is a minimum or even an extinction, there is actually a maximum, namely that of the sideways motion of the water molecules.

If we don't look at the whole thing through the isolated view of our teachers' glasses, but look at the dynamics of the water molecules or the energy distribution, then these properties are distributed completely homogeneously over the surface or the medium of the wave propagation, despite the interference. An extinction only takes place for isolated observation features. In the sum of the features, however, the dynamic distribution is also completely homogeneous in the case of interference.

What about the double-slit experiment with the light? Here, too, amplification and extinction are spoken of.[12] p.167 Fig. 5.19 But what can actually be determined.

First we send monochromatic light from a laser onto a screen. In between we put a filter with which you can darken the beam further and further. The more you darken the light beam, the more clearly you can see that individual light photons are arriving on the observation screen. Here there is no completely homogeneous distribution on the screen. If the water waves are generated with less and less energy, they become flatter and flatter, but the dynamics remain completely homogeneous over the surface.

In the case of light, one can statistically determine an even distribution of the photons over the surface of the screen. Which can then also be described as homogeneous for the mean value. And this distribution can perhaps also be described with the same formulas, but there is an actual difference in the distribution.

There is nothing next to the point where a photon was measured. In contrast, the wave is continuously distributed in space. Even if you go down to the atomic level with the water wave, you could describe the space between the molecules as empty. But as long as I can still observe a wave motion, there is a fairly 100% probability that a molecule will move past here as part of the wave motion, or I no longer have an observable wave. Whereby the motion of the molecule also corresponds to the wave motion and not only to the Brownian molecular motion.

Let's go to the double slit experiment. With the water wave, we have maxima and minima when looking at the ups and downs in isolation.

But there are also smooth transitions in between. Even when looking at the characteristics in isolation, we don't suddenly have nothing next to the maxima, but smooth transitions to the minima. And these minima only affect the characteristic considered in isolation. There is a maximum at this point for sideways motion. So here only one feature is "erased". The energy or dynamics of the water molecules is the same at this point. With the photons we can also determine a line-shaped maximum, but there is nothing directly next to it. No transitions, no light of different frequency, no light with different polarization, no other radiation like radio waves and also no heating of the screen in which the energy could dissolve. There's just nothing here. Also, no extinction of an isolated feature (up and down) caused by the maximum of another feature (sideways motion).

For me, the double-slit experiment is a prime example of how light consists of particles and has no wave properties.

Waves are the motion of a medium, where the components of the medium must be significantly smaller than the appearance of the wave. The wave can represent changes in density, as in the case of sound, or vibrations, an interface, as in the case of a water wave.

As an example, a ball pit that anyone who has ever given their child to the play area of a department store will know. If we could design the surfaces of the spheres so well that they hardly have any friction, then we could use them to represent waves.

Particles are spatially delimited structures that can be traced through time. They can also have a fixed relationship to a medium, like mass particles have to the gravitational field. Which in any case can only move up to the speed of light. And as I assume it, light photons travelling as massless particles at the speed of light towards the gravitational field. I can't say what the gravitational field consists of. I don't want to finish examining this here either, but figuratively speaking I just want to examine the water waves first, without knowing what water is made of.

An attempt to visualize particle interference: Imagine two guns that can fire bullets in very rapid succession, so that the gaps between them are not much larger than the bullets. Now I set up the bullet rays so that the rays cross. Then I slide one of the guns back and forth. Once the balls just fly past each other. If I move the rifle by half the distance between the bullets, they collide and the picture that emerges is completely different. But just because you can represent something like interference with balls, they are still not waves.

However, this example should only be taken as an indication. M. Born[6] p.75 writes: "…who made Huygens a keen champion of wave theory. As the first and most important argument for this he saw the fact that two rays of light cross each other without affecting each other, just like two waves of water, while bundles of ejected particles should collide or at least interfere."

So you have to research the properties of the light photons more closely. Perhaps, as massless particles, they always need a mass particle to cause an interaction. That's why they don't influence each other.

Another example. Let's look at a drop of water floating in zero gravity. It can rotate and thus becomes elliptical, it can swing in the direction of its motion or crosswise to it. In this way, he can create wave patterns in his silhouette as he moves. For me, however, the drop of water remains a particle and not a wave. Light photons could also oscillate or circle in the extension of their effective range and thus produce e.g. the polarization. Nevertheless, it remains a particle.

What causes the light particles to move on such special paths after the double slit. - Now I get problems with the quantum physicists -. I think at the double slit there is an interaction or not. The interaction is a multiple of a certain size. Depending on the energy content of the photon itself, this results in a correspondingly strong deflection.

I would like to do research in this area, because from my point of view the interaction that causes the resulting image on the screen can only take place completely causally at the double slit. But you have to be willing to search here. Not with the mathematical glasses, but with the mental dissection knife, which then has to be implemented in experiments. Only the attempt can help us to get closer to the reality of the universe around us.

### *3.9 Radio Waves*

How about the radio waves? Do they show particle character at any point? Certainly they are generated by the oscillation of electrically charged particles and in turn they cause electrically charged particles to oscillate. But at what point do the radio waves show particle character? Light photons can be used to shoot atoms, which then change their motion at a very specific angle and at a very specific speed. This change is fully consistent with the energy content of the photon. Can you do something like this with radio waves? In particular, could you change the moving of just one of a group of electrons?

From my point of view, radio waves are oscillations of the electromagnetic field. Whatever it is, it has nothing to do with the gravitational field or the motion of light.

Mathematically, these waves can be described in exactly the same way as light. But contrary to what is commonly said, one cannot conclude that light photons and radio waves have the same physical basis. Here, too, I would like to try to look out of the box with suitable experiments.

Above, figuratively speaking, I wasn't looking through the mathematical glasses, but looking out the window of the box to see if I was standing firmly on the ground of facts, or moving through space at a steady pace. I found out that light photons do not have wave properties in the sense of an oscillating medium and that radio waves do not appear to be spatially delimited particles. The electromagnetic field could possibly consist of small particles, like water consists of molecules. However, these do not appear individually in the wave structure.

I would like to come back to the radio waves of the electromagnetic field when it comes to the method of measuring the motion against the gravitational field. More precisely, to be able to do this.

### 3.10 Coincidence, does it exist?

Everyone knows games with dice or roulette that are supposed to be based on chance. But are these really random? In roulette, the croupiers are changed regularly. Resourceful people have found out that the croupiers throw in such a way that after one of the numbers there is a high probability that the next time the ball will land in a certain area of the disc head, these numbers are more likely to appear than numbers from another area of the disc head.

How can that be if the process was really random? From this one can conclude that the process is not accidental at all, but quite causal. Only the many touches cannot be described so precisely that the result could be predicted exactly. As I already wrote in the chapter about natural reality.

What about radioactive decay? It's supposed to be random, because you can't predict when an atom will decay.

Let's imagine a vibrating plate with a rim on which we want white and red balls to be. On the board we paint one or more fields of different sizes. The plate is observed by a camera with a computer, like a camera on a self-driving car. And whenever it detects exactly two white and two red balls in one of the fields, the plate is switched off.

As in billiards, the balls move causally on the table and the turning off is also causal. Now you set up a million such plates of the same type and with the same fields and start. I'm sure that you could also determine a half-life until all plates are switched off. Depending on how many balls are on the plate and how the painted fields are designed, this half-life will be very different.

We can now paint other fields. And whenever there is exactly one red ball in the field, the plate is switched off. Depending on the number of balls and the size of the fields, this will also lead to a half-life until all plates are switched off.

For me, both are completely causal processes. Why shouldn't something like this also happen in atomic nuclei. The first analogue to alpha decay, the second analogue to beta decay. I don't think anyone can say they know exactly what happens in atomic nuclei. But based on this analogy, I believe that radioactive decay is also a causal process, only the observing person cannot predict this exactly, but has to be satisfied with a statistical statement.

But there is something else in this universe. Let's imagine a meteorologist would be so far with his predictions that he could mark an area on the ground and say exactly where the first raindrop would fall. Then it's time and I come, hold my hand in between and catch the drop. So his prediction is wrong. The drop did not fall in the predicted place.

I'm only human and can only perceive the universe that surrounds us with my senses and only have the knowledge available that other people have already collected before me. But maybe there is also something like a hand in inanimate nature that doesn't let things happen the way they causally should.

But wasn't the action of my hand causal? I wanted to show that, despite all the precision, the meteorologist cannot predict the future exactly. Wasn't that causal now?

Whether the human will to act is free or causally predetermined is not the subject of this book, but the title is called The Relativity of the Observer and the observers here on earth are us humans. As we grow up, we accept things that become dogma for most people. Experience shows that it is possible to build up stable and successful structures in this way. But in basic research, one should be much clearer on the one hand, what is an actual observation and what is a conclusion that is actually correct only to the extent that the assumptions that you have to make for this conclusion are actually correct. In basic research, such assumptions must not become dogmas that can no longer be questioned and, if they are, answered with silence.

## 3.11 Equating observations and their measurements

The equating of values can be very dangerous in the scientific analysis of event scenarios and their mathematical derivation and lead to completely wrong statements. Things are always equated in the approach and calculation or left out because they would have no effect.

As an example, consider relativistic effects. The term square root of $(1-v^2/c^2)$ appears in the formulas. The value for $v^2/c^2 = 0$ is set for low speeds. If I only observe speeds up to 1000 km/s, which is quite fast, then the value for $v^2/c^2 = 0.000011$.

If I only collect readings up to the 3rd digit after the decimal point, then this value is below the measuring accuracy and can actually be neglected. But if I increase the measurement accuracy, then the value can gain meaning again, so you shouldn't forget that completely. But sometimes it's not just about values that differ so much in a certain range that they fall below the measurement accuracy.

To explain it visually, we take 2 ladders, one of which is 10 m long and the other only 5 m long. No matter how many times I halve the length of the ladders and if one is only 0.000,000,001 mm, the other is still half that size. No matter how much the values are reduced, this does not lead to a change in the basic statement.

An example of a transition area: [12] p.516 *"Every space, no matter how curved, is locally (i.e. on sufficiently small areas) flat."* So for a sufficiently small area, the spherical surface is flat.

On a spherical surface, the triangle can correspond to a maximum of one degree of longitude. Then the angles would be 3 x 180 degrees. The smaller we choose the range, the value will be smaller very quickly. In the case of a very small area, it is actually almost only 180 degrees, but only if at least one of the sides only had the length 0, then it could correspond to 180 degrees. But what is the angle at the end of a geodesic.

As soon as it is still a triangle, the sum of the angles differs from 180 degrees somewhere after the decimal point. Here we are with the measurement accuracy. If the difference is only after the decimal point, at one point is smaller than the measurement accuracy, it doesn't matter. But you have to stick to the considerations in this small area.

Here is the ladder example. Let's take 2 ladders. One is 1 m long, with a ratio of 100:1 to the other ladder, which is only 1 cm long. Now the long ladder should only have a length of 1 mm. However, it is only measured up to 0.1 mm. Then we extend the ladders 1000 times again. Then we would again have a ladder of 1 m, but the other one would have disappeared, although here it should have a length of 1 cm. During this arithmetic operation, something suddenly disappeared. So we have a completely different situation, even if we calculated correctly. Such transitions must not exist in arithmetic operations.

Let's talk about statements like:

*"So can we actually call a laboratory on Earth an inertial system under these circumstances? - Indeed, we can! The reason is that all these motions deviate from a uniform motion only in an extremely small degree. So small that we can hardly measure the effects even in a laboratory with the most sensitive devices."*[12] p.71

With today's precision of atomic clocks, however, we can already measure the difference between a spatial synchronization of the clocks according to Universal Time Coordinated or Einstein's definition of simultaneity. At the equator it is 206 ns for a ring without gravitational effects for the circumference of the earth. However, for a distance of 21 km it is only 109 picoseconds, which corresponds to 1 cycle of the defined second. For one meter it would only be $5.2 * 10^{-15}$ seconds. It becomes difficult with the measurement accuracy if we want to measure it within a laboratory. As measurement accuracy increases, we need to keep that in mind.

The extent of the difference is determined exponentially by the speed. By reducing the distance, however, it only decreases linearly and the ratio remains the same. It is therefore only dependent on the measurement accuracy, it will always retain its magnitude, no matter how small the section is.

Since the clocks on the earth's surface actually work differently, there must also be an actual contraction in length. Therefore a laboratory rotating with the earth is never an inertial system.

# 4. Minkowski diagrams[22]

*Note on the graphics. Some readers only display these insufficiently and the explanations contain characters that are also not displayed correctly. The characters are described with the preceding words so that they can also be understood with a black and white reader. If you send me an e-mail to buch@darmer.de with a copy of your invoice for the book, I will send you the drawings as a PDF file.*

## 4.1 Basic Terms

In the following I would like to present problems for which I have not yet found a comprehensible description, but which can be described very clearly. This can also help to understand the Lorentz transformations. First, some words like "seeing" or "measuring" need to be explained in more detail.

I would like to explain this using this problem: How do two observers moving past each other at 0.6 times the speed of light see a flash after 1 second (1 s) that they emitted together when they met?

A **flash** is a light signal that, after it has been triggered, generates world lines that spread out in all directions and as such a signal can also be clearly tracked spatially and temporally, as I explained in chapter 2.8 about linearity.

The speed of light and the speed of the observers should not be a problem. The 1 s must be measured by both observers in their own time.

Seeing means that information signals in the form of light from different events are perceived by one eye at the same time. Of course, this can also be a measuring instrument. However, the simultaneous arrival says nothing about the times at which the light signals were generated. If an observer sees something within a system in which the observers are at rest with respect to each other, all other observers can, through the spatial synchronization of their clocks, assign this event to a proper time, namely the same time displayed on their clock, even if the observers have not actually "seen" the event at that time. It was then measured within the system or measurement arrangement. The correctness of the temporal assignment of the measurement depends on the correctness of the spatial synchronization of the clocks.

In the following, I would like to represent such an **observer** as $B_0()$. B is an observer who does not move to the other B observers. This means a time signal from a B-observer e.g. $B_0()$ sent out regularly, at another, e.g. $B_1()$ reflected and received again at $B_0()$ leads to a constant time difference. The B observers are named after half the value of this time difference. This means that $B_1()$ is 1 light second (Ls) away from $B_0()$. The time value to be read off the watch of the observer is written in brackets. These time values are arbitrary definitions, only the frequency of the oscillation of the caesium atom of the atomic clock is fixed by nature. A time signal sent out by $B_0(0)$, which is back at $B_0(2)$, was thus reflected at $B_1()$. According to Einstein's definition of simultaneity, the observer $B_1()$ must set half the time for the outward and return journey to the event of reflection, i.e. half the time difference at $B_0()$. In this case half of 2, i.e. 1. There is therefore an event $B_1(1)$ which would be spatially simultaneous with the event $B_0(1)$ according to Einstein's definition of simultaneity. Since these events can only happen in one place and at one time, they are two **one-place-one-time events**. These are therefore valid for all observers of this universe. There must be event $B_0(0)$ for all of them, in which the time signal 0 is sent out, which arrives there at event $B_1(1)$ and is back there at event $B_0(2)$.

Depending on the position of the other observers in the B-system, or whether they are moving towards it, the light travels for different lengths. This means that what the observer sees at the same time only happened at the same time in very specific cases. An observer $B_{-1}()$ located on the other side of $B_0()$, also 1 Ls away as $B_1()$, would receive the information signals from events $B_1(1)$ and $B_0(2)$ simultaneously. To make it clear that these are information signals and not the events themselves, I represent them as $[B_1(1)]$ and $[B_0(2)]$. But since $B_1()$ is further away from $B_{-1}()$ as $B_0()$, the event $B_1(1)$ must also have taken place before the event $B_0(2)$ for it.

With $B_0()$ there is the event $B_0(2) + [B_1(1)]$. The information signal from this $[B_0(2) + B_1(1)]$ can send $B_0()$ in all directions. In principle, every observer in this universe can receive this signal. Therefore, for causal reasons, there cannot be anyone for whom event $B_0(2)$ occurred before event $B_1(1)$. In principle, one could do the same with signals faster than the speed of light.

Unfortunately, the following sections can no longer be presented without numerical examples. I tried to keep them simple. However, it is in the nature of matter that it is no longer that easy, but with a little geometrical understanding you will be able to read into the Minkowski diagrams.


## 4.2 The flash after one second

If two observers emit a flash of light when they meet, it should form a sphere around the respective observer in three-dimensional space after 1 s, since the speed of light should be the same in all directions. This applies to both observers, since they are equal in terms of motion and the measurement of the speed of light. But since the observers are moving towards each other, they cannot both be in the centre of the same sphere after 1 s.

Here's the problem, how to measure the **location** where the flash is after 1 s of proper time? The observer $B_0(0)$ can set up mirrors at a distance of 1 light second (Ls) on which the light is reflected, and if, after 2 s of its own time, the flash hits him $B_0(2)$ again from all directions at the same time, he can determine that that the flash needed exactly 2 s for the outward and return journey in all directions. He can also say with certainty that the events of reflection all took place after sending out and before receiving again. Everything else depends on assumptions or definitions. The speed of light is always measured at the same speed for the sum of the outward and return journeys together, a fact of observation. According to **Einstein's definition of simultaneity**[10], the light should take just as long to travel there as to travel back. By definition, the reflection happened exactly half the time of 2 s, i.e. after 1 s. If you set up clocks at the mirrors and set them to 1 s at the time of reflection, then they are spatially synchronized with each other according to Einstein's definition of simultaneity. If you use ideal clocks[17], they are also synchronous. **Synchronous motion of the clocks** means that they always show the same time value when they are in a common place, or e.g. with several clocks that represent different time zones of the earth, then always display the same time differences. If the clocks are in the same place, comparing the times is not a problem. If they are separated from each other, but the distance to each other is constant, then a time signal from one clock always arrives at the other clock with a constant difference.

The whole thing is represented in Minkowski diagrams. Initially only for the B system and only in one spatial dimension, which is represented by the X-axis. The two observers in opposite directions on the sides of $B_0()$, each 1 Ls apart, we call $B_1()$ and $B_{-1}()$.

The three events marked with a black dot ● in **Figure 1** correspond to $B_{-1}(1)$, $B_0(1)$ and $B_1(1)$. In the B-system, these are the events that are simultaneous at time $t = 1$. At time $t = 2$, the lightning reflections all arrive again at $B_0(2)$, marked with a black box ■. In the lower part of Figure 1, the point in time $t = 1$ is viewed from above in 2 spatial dimensions, precisely from the direction of the t-axis. The lateral events at a distance of 1 s in the area are also shown here and result in a circle at time $t = 1$.

---

17  Einstein-Langevin clock [17] S. 25

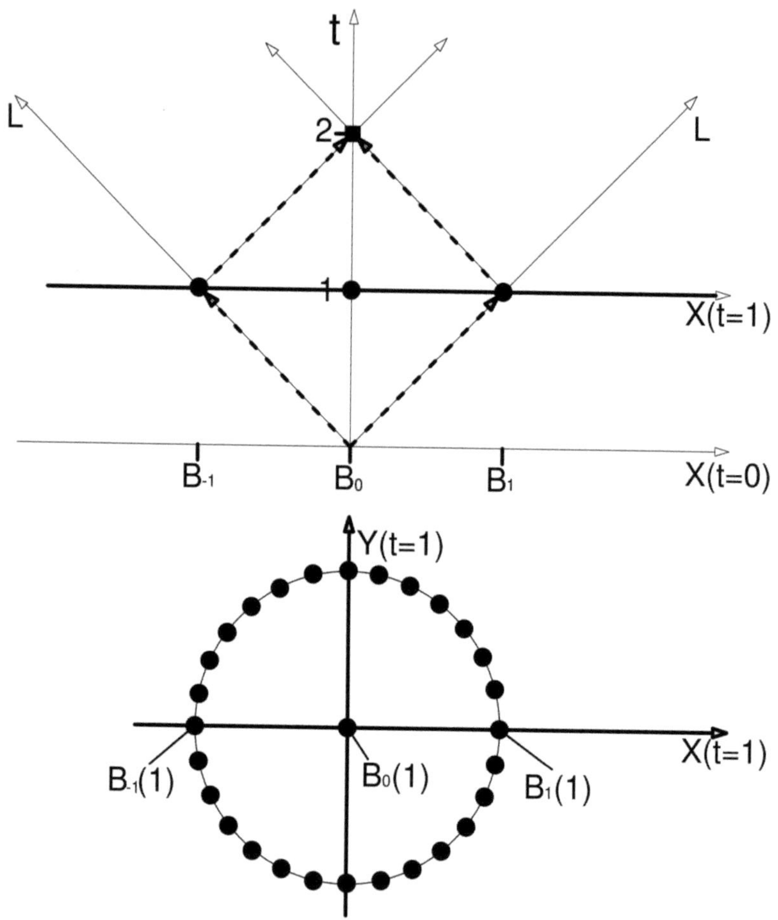


**Figure 1:** One second flash in the B-system.

Now the mapping is completed by the system of the observer $C_0()$. At rest, he is said to have set up the mirrors around himself just like $B_0()$. Now he is moving past $B_0()$ at 0.6 c and the moment they meet they want their clocks to be set to 0. So there is the event $B_0(0) + C_0(0)$. After measuring the B-system, a length contraction corresponding to the Lorentz transformation occurs in the C-system. So at time $t = 0$ he measures the observer $C_1()$ at a distance of only 0.8 light seconds and also the observer $C_{-1}()$ in the opposite direction. For $B_0(0)$ there are the events at time t=0: $B_{-1}(0)$, $B_{-0.8}(0) + C_{-1}()$, $B_0(0) + C_0(0)$, $B_{0.8}(0) + C_1()$, and $B_1(0)$. In **Figure 2**, these events are marked with a black dot ●. The motion of $C_0()$ through the B-system is represented by the time axis t'. After 1 s of the $B_0()$ time, the $C_0()$ moving at 0.6 c is at $B_{0.6}(1)$. Since there is a time dilation at $C_0()$ from the point of view of $B_0()$, the clock of $C_0()$ at time t = 1 s only shows t' = 0.8 s for this event. For $B_0(1)$ there is a simultaneous event $B_{0.6}(1) + C_0(0.8)$. At the same time, the events $B_{-0.2}(1) + C_{-1}()$ and $B_{1.4}(1) + C_1()$ are also marked in the B-system with a black ring ○ in the drawing.

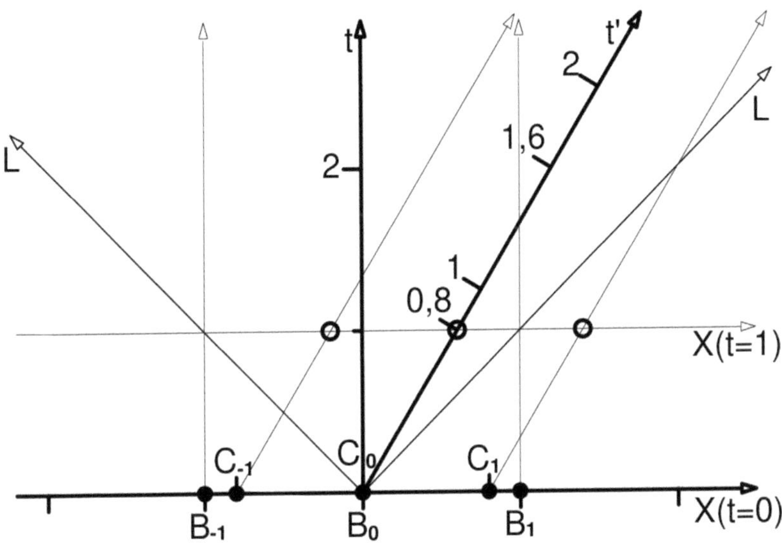


**Figure 2:** C-System is inserted.

Now let's reflect the lightning from the point of view of the B-system at the mirrors of the observers $C_{-1}()$ and $C_1()$. The events are marked in **Figure 3** with a dot ●. Then the lightning reflections measured by $B_0()$ arrive again at time t = 2.5 at $C_0()$, marked with a box ■ in Figure 3. $C_0()$ is then located at $B_{1.5}(2.5)$ measured from $B_0(2.5)$. After the Lorentz transformation, the clock of $C_0()$ must indicate the time t' = 2 at that point in time. So there is the event $B_{1.5}(2.5) + C_0(2)$ in which the lightning bolts reflected in the C-system arrive at $C_0()$ simultaneously for all observers. For the spatial simultaneity in the C-system, according to Einstein's definition of simultaneity, the clocks for the lightning reflections at $C_{-1}()$ and $C_1()$ must indicate half the time for the outward and return journey that was measured in the own system, i.e.

half of 2 s. They must therefore set their clocks to the time t' = 1 for these events.

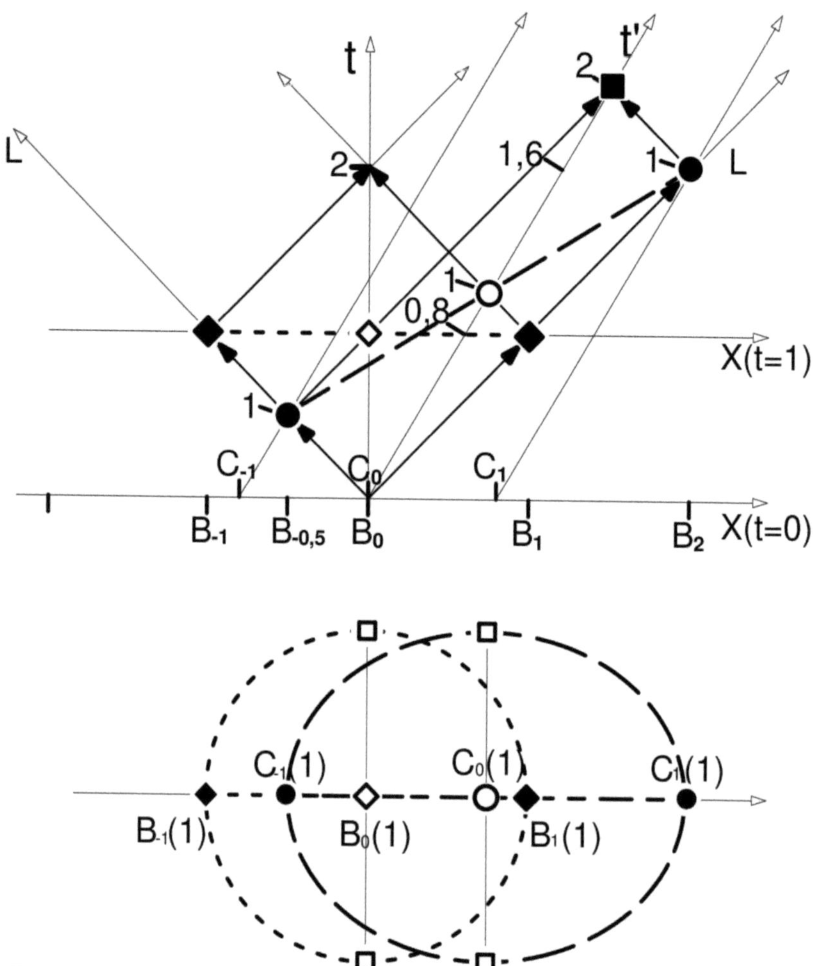



**Figure 3:** One-second flash in the C system from the perspective of the B system.

The events $C_{-1}(1)$ and $C_1(1)$ are at the points marked with a dot ● in Figure 3. Viewed from the C-system, these events are at the same time as event $C_0(1)$ marked with a circle ○. Since there is no lateral contraction in length, viewed from above, the mirrors set up laterally by $C_0()$ in the C-system are just as far away from $C_0()$ as the mirrors set up laterally by $B_0()$ in the lower part of the Figure 3 each marked with a

68

square $\square$. Viewed from above, the lightning reflection events measured from the B-system form a circle in the simultaneity plane of $B_0()$ and the reflections from the C-system mirrors form an ellipse. These figures correspond to the sections through the light cone resulting from the simultaneity planes of $B_0(1)$ and $C_0(1)$. However, the events of the C-system are not measured as simultaneous events in the B-system. The reflections from the C-system mirrors, which are inside the circle in Figure 3, are measured in the B-system at time $t = 1$ as having already happened, and those outside the circle as not yet having happened. The event $B_0(1)$ was marked with a diamond $\diamondsuit$. The events $B_{-1}(1)$ and $B_1(1)$, which are spatially simultaneous for him, are marked with a rotated box $\blacklozenge$. First, in the upper part of Figure 3, further one-place-one-time events are marked with circles $\mathbf{O}$ and diamonds $\diamondsuit$, shown in **Figure 4a**. The diamonds $\diamondsuit$ mark events on the simultaneity levels of the B-system and the circles $\mathbf{O}$ mark events on the simultaneity levels of the C-system. All events were measured from the B-system. Since they are all one-place-one-time events, they must also be measured from the C-system when it considers itself at rest. To do this, we set the t'-axis and x'-axis at a 90 degree angle, shown in the upper part of **Figure 4b**. For $C_0()$ the whole thing is exactly the opposite. Purely due to the geometry of the Lorentz transformations and the exclusive use of ideal clocks and light signals, the observers $B_0()$ and $C_0()$ are completely equal, taking into account Einstein's definition of simultaneity.

Even if you insert an observer $D_0()$, which moves in the opposite direction to $C_0()$, there are no contradictory results from $C_0()$'s point of view either. From the metrological point of view of the C-system, the 1 s light flash circle measured by $D_0()$ would only have been more oval, corresponding to the even steeper cut through the light flash cone.

How is it that, despite the images that look so different, there are no causal contradictions? No physical changes are made for either approach. Each observer with his clock forming a common world line, their encounters with the observers moving towards them and their flashes of light or time signals form a web of one-place-one-time events. Depending on which observer system considers itself to be at rest, this net is pulled in one direction or the other like a fishing net. However, no node is resolved and no new nodes are added. That's why this picture remains free of contradictions.

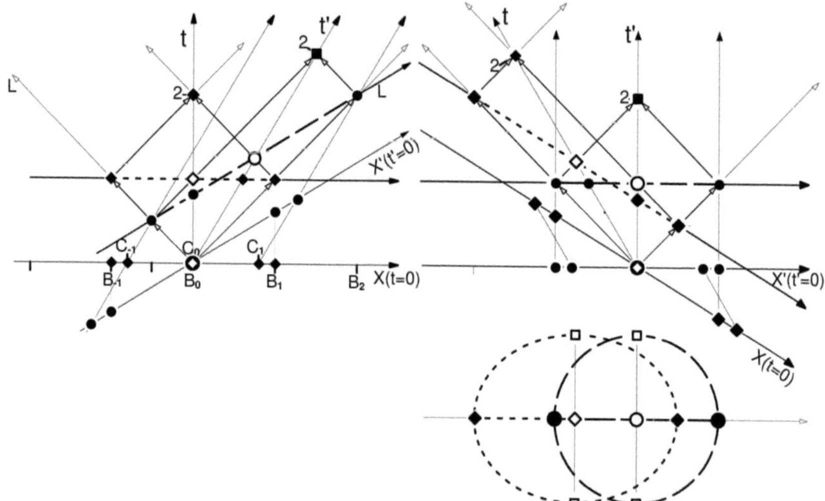

**Figure 4a and 4b:** All one-place-one-time events also measured in the C-system.

It is difficult when one of the two observer systems turns around and they move past each other again. For this at least one of the two observer systems must get a kink in their world lines. To do this, we just let the twin brother of $C_0()$, we call it $D_0()$, fly back. To do this, he has to expend energy so that he changes his speed in relation to the other observers in his system and thus also in relation to all other observers. If you look beyond the edge of the drawing, it also changes its state of motion in relation to the entire universe. Systems B and C measure its speed differently. If one calculates the rate of the clock of $D_0()$ from the B-system from the Lorentz transformation, exactly the same values result that are also calculated from the C-system. From both of them, the same amount less time has elapsed when $D_0()$ meets $C_0()$ on $D_0()$'s clock. Better presented in Chap. 4.6.

Up to this point, all observer systems and observers can consider themselves at rest and measure exactly the same when viewing the other systems. You can also say that they are absolutely equal when it comes to determining the measured values. But what happens when the whole C-system wants to change the motion direction? As a simpler example, just want the twin brothers of $C_0()$ and $C_1()$ to reverse together. Geometrically, this no longer works on an equal footing. Let's assume that $C_1()$ fires the engines at event $C_1(1)$ in order to be able to return to $B_{0,8}()$. Suppose not only the information but also the mechanical pressure would propagate to $C_0()$ at the speed of light, then $C_0()$ would receive the information from the start at the event $C_0(2)$ and also feel the thrust at the same time. Now we let the C-system accelerate until it moves back through the B-system at 0.6 c.

If the acceleration is only small, then no relevant effect would result from the different propagation of the acceleration motion from $C_1()$ to $C_0()$. The different motions of $C_0()$ and $C_1()$, which from the point of view of the B-system result from the dissolving length contraction and the newly developing length contraction, would also have no significant effect.

From the point of view of the B-system, the C clocks run slower. As the speed decreases, this becomes less and less until, from the point of view of the B-system, they go at the same speed when they come to a standstill. When the speed increases in the other direction, they then go increasingly slower again. In the whole process, an increasing time difference develops between the clocks of the B-system and the C-system. But the time difference between the clocks of C0() and C1() would remain the same.

From the point of view of the B-system, the signal from $C_1()$ arrives at $C_0()$ much later and the signal from $C_0()$ at $C_1()$ much earlier. This is also measured in the C-system with the same setting of the clocks. Thus, in the C-system, the speed of light is no longer measured as c in both directions. The clocks in the C-system are also no longer synchronized with each other according to Einstein's definition of simultaneity. They have to spatially resynchronize their clocks after acceleration. To do this, $C_0()$ must put its clock back or $C_1()$ put its clock forward. For the new synchronization, the C-system then measures the speed of light again in both directions with c and the clocks from the B-system also run slower again compared to this spatial synchronization.

Is it permitted to adjust the measuring instruments during an experiment? If you recalibrate the measuring instruments to something else during a test run, you can hardly get comparable measurement results. Are the two parts of the experiment still comparable? The world lines of the caesium atoms of the atomic clocks do not jump during the course of the experiment. The jump is caused by changing a defined value, namely changing the displayed time. If the spatial synchronization of the C-system clocks were not changed, they would also measure the B-system clocks as running faster from this synchronization. So fast that the slower pace is overtaken by the first phase so much that the clocks of $B_0()$ and $B_1()$ each time $C_0()$ and $B_0()$ and $C_1()$ and $B_{0.8}()$ show exactly the amount that results from the Lorentz transformation from the point of view of the B-system.

I have described the effects of maintaining synchronization or adjusting the clocks in Chapter 4.6 on the twin paradox and Chapter 5.4 on measuring the motion to the gravitational field in more detail.

### *4.3 Faster than the speed of light c*

Anyone who thinks something like warp drives and wormholes is possible should read on here and not take the following as nonsense. It makes no causal difference whether I send a messenger back and forth with a warp drive rocket with my time signals or discover a new radiation that is propagating at a speed higher than the speed of light. The spatial simultaneity, which an observer cannot further limit in terms of measurement technology, is determined by the greatest possible speed with which information can be passed on. Also, the arrival of the warp-powered messenger can only be after its departure and must be causally before its return to the starting point if it has travelled back. In principle, this does not differ from a time signal that was sent faster than the speed of light.

Suppose radiation was discovered travelling at four times the speed of light. Then an observer $B_0()$ should be able to set up receivers $B_2()$ and $B_{-2}()$ two light-seconds away, which receive the signal and send it back, similar to the mirrors in a flash of light. These events are marked in **Figure 5** with dots ●. Then the answer should be back for him from both directions after a second, marked with a circle O. These signals could also be reflected from the B-system when they reach observers $C_{-1}()$ and $C_1()$, marked with boxes ■. Then these reflections at $C_0()$ would also be back at the same time, marked with a square □. Since these events are one-place-one-time events, they must also be measured in the C-system. $C_0()$ might determine that the signal would be back before half a second elapsed. Since $C_1()$ and $C_{-1}()$ are each measured a light second apart in the C-system, he measures the signal back faster than four times the speed of light.

Even more amazing would be the arrival of the signal at $C_1()$. Sent out at time $t' = 0$ at event $B_0(0) + C_0(0)$, the signal arrives at $C1(<0)$ well before its clock indicates time $t' = 0$, i.e. before event $C_1(0)$ (marked as a diamond ◇). In the spatial synchronization of the C-system clocks, the arrival of the signal at $C_1()$ is already measured before it is sent out. Nevertheless, the signal for the way back takes so long that the answer is only back in the future. This applies in reverse order to the signal that is sent back at $C_{-1}()$. It arrives at $C_{-1}()$ well after $C_{-1}(0,5)$ and is then back at $C_0()$ before $C_0(0,4)$.

I had already written that the signals should be sent back by a B observer. Because based on any laws of conservation of energy and the absolute equality of the inertial systems, from his point of view $C_0()$ should also be able to emit the radiation. First, in **Figure 6a**, the event sequence is shown from $C_0()$'s point of view, keeping all one-place-one-time events, and in **Figure 6b** then the radiation emitted by $C_0()$ is shown as thick black lines. But what kind of strange radiation would that be? It is emitted simultaneously by two observers moving towards one another when they meet at a point in space. This corresponds to a one-place-one-time event. The signals then move completely differently in space and time (represented by thick arrows), without regard to any causality or conservation of energy.

I'm not willing to allow breaches of energy conservation or causality. I also do not believe that there is such a strange radiation that does not follow a uniform principle of motion, such as sound, the motion of masses or the light-speed signals do, each of which follows a uniform principle of motion.

Suppose we discovered a way to send information signals faster than the speed of light. Then it will spread out symmetrically in all directions according to a motion principle. For an observer $B_0()$ who is at rest with respect to this principle of motion, the picture could then look like that in Figure 5. The arrows shown in the figure for four times the speed of light also correspond to **world lines**. For reasons of causality, these must also apply to all observers. Therefore, also for the C-system, the sending of the signal at $B_0(0) + C_0(0)$ must be before the event of the arrival at $C_1()$. For this event, marked with the right box ■ in Figure 5 and 6, a value $x > 0$ must be set on the clock of $C_1(x)$. For the C-system, with the clocks set according to Einstein's definition of simultaneity, the only conclusion left for causal reasons is that this spatial simultaneity does not actually exist.The world line of the clock $C_1()$ and the oscillation of the caesium atom are not affected by this. Only the time value set arbitrarily on the watch is incompatible with this. According to Einstein's definition of simultaneity, the clock is causally wrong. Thus, for him, the speed of light cannot actually be the same in all directions.

In chap. 5.4 I describe what the situation looks like during rotation. Here, the spatial simultaneity can be more causally restricted by measurements other than ideal clocks and the emission of light signals by the observers directly to one another. For this reason it is not possible to measure the same speed for all observers in both directions of a tangential motion. In the Universal Time Coordinated, the speed of light is therefore not measured at the same speed in east and west directions.

For the exact straight-line motion it remains: either, there is no causally higher speed than the speed of light, no matter whether due to the quantum effects, wormholes, tunnel effects or gravitational waves, or K and K' are physically not completely equivalent.

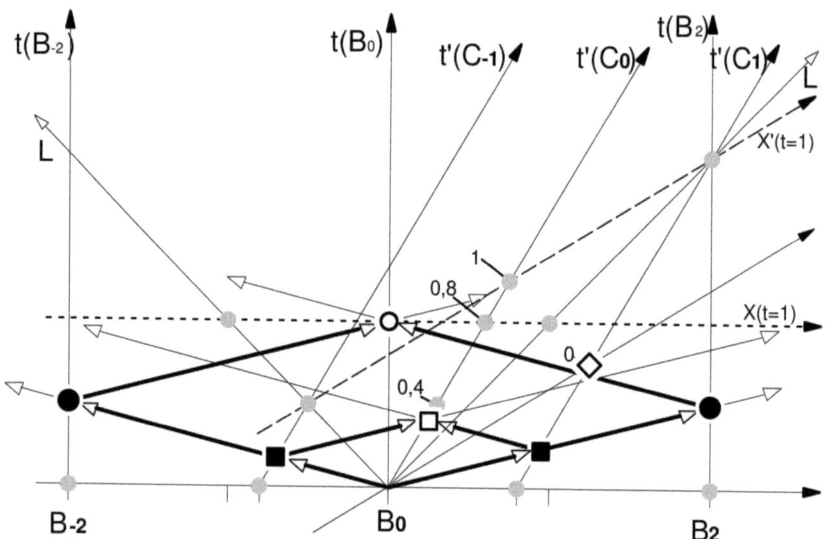


**Figure 5:** Information signal 4 c fast.

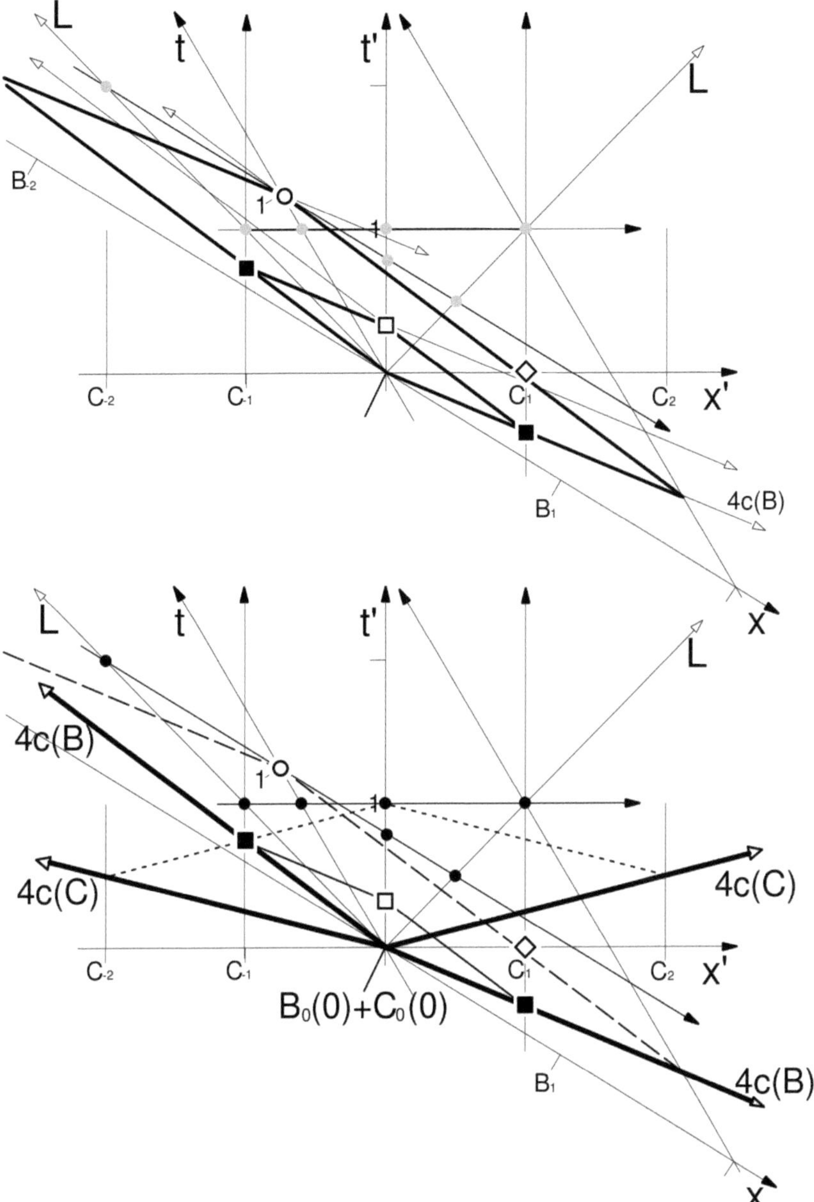



**Figure 6a and 6b:** 4 c fast information signal from the point of view of the C-system.

## 4.4 *Transfer to satellite navigation and Universal Time Coordinated (UTC)*

I would like to apply the problem to a specific situation. Let's imagine a part of the equator or a part of the earth's orbit around the sun. Or better yet, a section on a lightweight ring that should correspond to the equator and place it in a void, far away from gravitational masses. Atomic clocks are placed on the rings.

The B-system represents the stationary ring in which no centrifugal force is measured. The C-system would correspond to the rotation at the equator. In another case, it could also correspond to the motion of the earth around the sun. With the rings, however, one could also imagine any other size of the ring or the speed. In order to keep the whole thing understandable, I will only present the relative proportions here.

On Earth and in the GPS satellites, clocks are all trimmed to Universal Time Coordinated UTC. This means that the rates of the atomic clocks are set in such a way that they are synchronized to a unit time despite different gravitational fields[18] and different motions[19]. And the clocks stationary on earth are also spatially synchronized to a unit simultaneity. However, this does not agree with Einstein's definition of simultaneity.

Some physicists say considering the Sagnac effect is trivial. But it is not at all. Such a quantity could also be introduced for the inertial systems and all would have a uniform spatial simultaneity. Then there would no longer be any difference for the spatial synchronization of clocks that move along the edge of a rotating system and the clocks that move tangentially to it. No matter what now causes the rotation effect, there would then no longer be an absolute constancy of the speed of light for straight-line motion either.

Again, for clarity, let's take the example where $C_0()$ moves past $B_0()$ at 0.6 c. We develop the whole thing from the point of view of $B_0()$, i.e. the way $B_0()$ would measure it. If we again use the rings far away from other masses, then the rates of all clocks can be set according to the definition of the second. Differences in gravity need not be taken into account, and all clocks move in the same orbit. Consistent with today's

---

18 Depending on the gravitational field, clocks run faster on a mountain than lower down at sea level. Aside from the altitude, there are also regional differences.

19 The GPS satellites move with relevant speed to the earth, or rather to the non-rotating reference system. In addition, the clocks on Earth also move at different speeds, depending on the degree of latitude at which they are located.

generally accepted views, the rotation leads to an actual time dilatation and thus, logically, to an actual length contraction. So the rate of the clocks slows down, and the standard meter gets shorter the faster they move towards the non-rotating ring.

To simplify the picture, let the observers $B_X()$ measure a circumference of 1 light second (1 Ls) for their ring, which is not rotating. From $B_0()$ the observer $B_1()$ with a mirror is set up in 1 Ls to the east and the observer $B_{-1}()$ with a mirror to the west in 1 Ls. Since the ring has a perimeter of 1 Ls, all three observers are in the same place on the ring. **See Figure 7.**

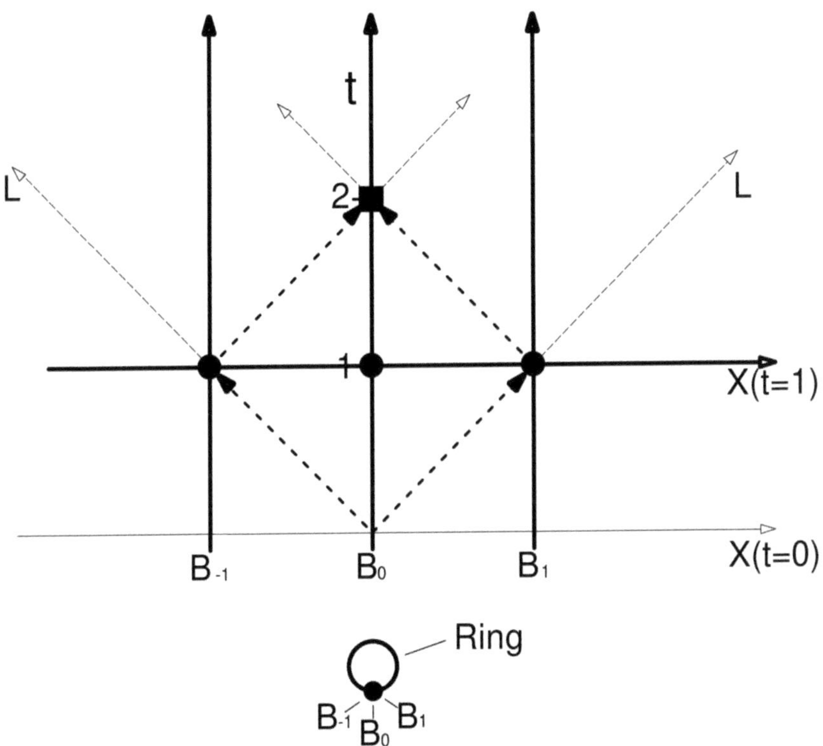


**Figure 7:** Ring at rest

Their world lines thus run on one axis: $t_{(B0)} = t_{(B1)} = t_{(B-1)}$. If $B_0()$ now emits a flash of light, according to the clocks of all three observers, it arrives simultaneously at all three observers from east and west after 1 s (marked by dots ●) and the reflections arrive again after a total of 2 s to all three observers (marked by boxes ■). The difference from linear or inertial motion is that the three observers on a straight rocket could not directly compare their clocks. Viewed from above, the X-axis shown in the lower part of Figure 7 is rolled up on the ring in such a way that the t-axes of $B_0()$, $B_1()$ and $B_{-1}()$ lie at one point on the ring. The world lines of the three observers are identical, so their experienced sequences of events are identical. In the upper part of the figure, the three t-axes represent the same world line. On the non-rotating B-ring, spatial synchronization also corresponds to Einstein's definition of simultaneity.

What about the turning ring? This is where an actual length contraction occurs. At a speed of 0.6 c, the C observers have to lay out 25% more standard meters in order to achieve the same ring size that corresponds to the B system. If the C-system corresponded to the B-system at rest, the observers $C_0()$, $C_{1,25}()$ and $C_{-1,25}()$ now move on a common world line, comparable to the observers $B_0()$, $B_1()$ and $B_{-1}()$.

Now $C_0()$ and $B_0()$ emit the flash of light when they meet, i.e. at event    $B_0(0) + B_1(0) + B_{-1}(0) + C_0(0) + C_{1,25}(0) + C_{-1,25}(0)$,   which   in Figure 8a corresponds to three of the dots ● on the x-axis for $t = 0$ and the three dots ● in Figure 8b. It makes no sense to set different times on clocks that move next to each other and are supposed to show the same present time at this point in space. This is why the start time 0 was also set on clocks $C_{1,25}()$ and $C_{-1,25}()$. This would also correspond to the synchronization of the earth clocks in UTC.

The **top part of Figure 8** shows what the events look like from the perspective of the B system. The development of the flash for the B system is shown in solid lines. The lightning bolt is shown in rough dashed lines for the complete C-system from $C_{-1,25}()$ to $C_{1,25}()$ and in fine dashed lines for the shortened part of the C system from $C_{-1}()$ to $C_1()$.

What does the sequence of events look like from the point of view of the C-system? Under these conditions, the spatial simultaneity synchronization is identical for the stationary ring as for the rotating ring. The moving clocks of the C-Ring are slower. From the C-system's point of view, however, the resting clocks are also running faster here. The flash of light also needs in the C-system, according to its own clocks, for the outward and return journey together, exactly the time it should take according to the length of the standard meters laid out. Both for the entire C-system from $C_{-1,25}()$ to $C_{1,25}()$ and for the portion from $C_{-1}()$ to $C_1()$.

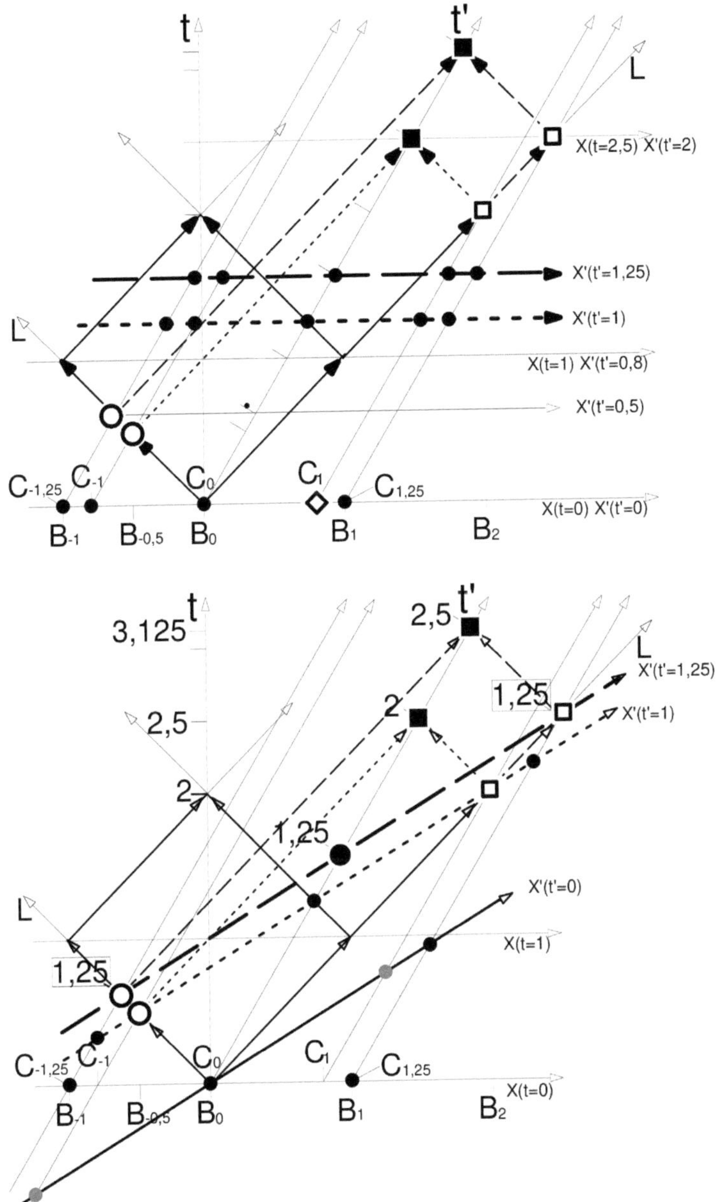



**Figure 8a:** UTC spatial synchronization
**8b:** according to Einstein's definition of simultaneity

The westward part of the light flash reached the group of observers around $C_0()$ after 0.5 s of their clocks (marked with the lateral circle ◉), the eastward part only after 2 s (marked with the lateral square ☐). The reflections from both directions are back at the same time after 2.5 s (marked with the upper box ■). Also for $C_0()$ the ring supplemented with standard meters to 1.25 Ls with light (half the time for the outward and return journey together) is measured as 1.25 Ls long. But for the entire circumference, the speed of light is not measured equally in both directions. This should then also apply to each section. It does not matter whether the diameter of the ring corresponds to the earth's orbit or e.g. three times the diameter of the Milky Way. There is no reason to assume that the basic proportions change with size alone. The corresponding synchronization times were also entered for $C_1()$ and $C_{-1}()$. The spatial synchronization in the moving C-system is parallel to that of the non-rotating B-system and thus also corresponds to the spatial synchronization of the clocks fixed on Earth in UTC.

Let us carry out a spatial synchronization of the clocks for the C system according to Einstein's definition of simultaneity. After that, the reflection events should be simultaneous, with half the time for the round trip. To do this, the clock at $C_{1.25}()$ must be set back by 0.75 s and the clock at $C_{-1.25}()$ must be advanced by 0.75 s. In the **lower part of Figure 8**, the time values are framed in boxes, which are created by adjusting the clock hands. The resulting level of simultaneity at $t' = 1.25$ s was shown in a rough dashed line. This has no influence on the world lines of the clocks and their rate. The differences in the time displayed over time, which reflect the clock rate, remain the same. The simultaneity level at time $t' = 1$ measured differently by adjusting the clocks in the C-system was also marked here by a fine dashed line to compare Figure 8 with Figure 3 (there the rough dashed line).

Since $C_0()$, $C_{1,25}()$ and $C_{-1,25}()$ share a common world line, they experience the one-place-one-time events marked with the lateral circle ◉ and square ☐, as 1.5 s events that are spaced apart and not simultaneous events. For causal reasons, they cannot spatially synchronize their clocks with each other in this way, but only as shown in the upper part of Figure 8. Thus the speed of light for the C-system is not equally fast in both directions. This should not only apply to the entire scope, but also to each section. The observers $C_1()$ and $C_{-1}()$ can no longer compare their clocks directly with each other, since they are not located at the same point in space. But even so, they have to fit into the synchronization so that their measurements match the measurements of the other observers, which also corresponds to the actual synchronization of the stationary earth clocks in UTC. Just as the earth clocks are already synchronized with each other for causal reasons, a signal with four times the speed of light (shown in Figure 5) would also lead to no causal problems. Under the real conditions of the earth clocks, the clock of $C_1()$ does not show the time $C_1(0)$ as shown in Figure 5 with the diamond $\Diamond$, but as shown in Figure 8a with the diamond $\Diamond$. With $C_1()$ and $C_{-1}()$, the arrival of the faster-than-light signal would definitely be after the transmission.

## *4.5 Even an actual effect is not always measured*

a) During rotation, there is an actual contraction in length. Nevertheless, the co-rotating observer cannot measure this as long as he is limited to a section. The observers $C_0()$ and $C_1()$ have a distance of 1 Ls at rest. If they accelerate to 0.6 c, there is an actual length contraction, which means that they are only measured for 0.8 Ls by the non-rotating observer. An actual time dilation occurs for their clocks, through both together the light signal for the outward and return journey together still needs 2 s for all observers on the clock of $C_0()$. The rotating observers use this to measure their distances (half the time for the outward and way back) still 1 Ls long. There would only have been occurred a gap in their ring.

If the pointer setting of the rotating clocks remains in a consistent synchronization, they remain in a simultaneity comparable to UTC. From the point of view of the rotating observer with this simultaneity, the ring at rest would have been lengthened, because after the acceleration it measures it to be 1.25 Ls long. If they had laid out their distance when not rotating with standard meters, nothing would have changed in the rotating state. They would measure the standard meter with their clocks and light signals further as 1 m, even if it has actually shortened. However, they would measure the still resting standard meters as extended to 1.25 m. Here again the question is: What has changed? Has the space itself changed, or does the standard meter actually have a different length under different physical conditions?

Comparable to the thermal expansion when heating a metal rod. If a metal rod elongates under the heat, then the space changes or my metal rod becomes longer. Here it is clearly decided that the metal rod will be longer because this solution is more compatible with other comparisons.

b) When idle, $C_0()$ emits a laser beam. This is shared as in the Michelson-Morley experiment. One part is sent to $C_1()$, reflected there at the mirror and then brought to interference with the other part at $C_0()$. This corresponds to a one-armed Michelson-Morley experiment, or extreme Kennedy-Thorndike experiment[17]. In principle, this also corresponds to a light clock.[21] p.91 We make the distance a bit more manageable, it should only be 1 micro light second, which would be about 300 m. Now the C-ring is slowly accelerated. Even if there is now an actual contraction in length and the clocks actually run slower, $C_0()$ does not detect a shift in the stripe pattern. He cannot measure the change in speed with this method.

If we mark the laser beam with time signals, the same effect would occur with $C_0()$. If these are reflected at $C_1()$, the time difference at $C_0()$ always remains the same. However, if $C_1()$ compares the arrival of the time signals with its own clock, it finds that the time signals are arriving later and later. If $C_1()$ caused $C_0()$'s laser beam to interfere with the beam of its own laser, then it would also notice a shift in the interference pattern. Apart from the increasing speed, the size of this effect depends solely on the distance between the observers $C_0()$ and $C_1()$. Since the effect for the outward and return path together always corresponds to zero, a light beam that is reflected back and forth a million times only leads to a million-fold zero effect. This should be taken into account when taking measurements with an experimental setup corresponding to the Michelson-Morley experiment. The MME also remains negative as a result of acceleration, even if there is an effect that can actually be measured.

Likewise, one could not determine with the one-way measurement whether the ring is rotating or not. This effect always occurs in the direction of acceleration, regardless of whether a rotation is accelerated in the same direction or an opposite rotation is decelerated. Expressed in general terms: An acceleration always leads to a later arrival of the time signals in the acceleration direction and an earlier arrival in the opposite direction. With ideal clocks and light signals alone, one cannot prove the rotational motion on small sections, there is no difference to the inertial systems moving in a straight line.


## 4.6 The twin paradox

So much has been written about it that I just want to mention a few points here. I was once recommended the book "Reisen durch die Raum-Zeit" by Leslie Marder [17]. If you have trouble with the twin paradox and don't trust me, you should read this because it's widely accepted by physicists. I only show some additions here. One thing is certain, although acceleration processes have an influence on the size of the time difference between the clocks, they have no fundamental influence on the tendency for which clock to show less elapsed time when they meet again. The only decisive factor for the effect is who changes the inertial system for the reunion, i.e. who turns around with expenditure of energy and flies back again. Reversing with the expenditure of energy not only means that it moves differently compared to the other twin, but it also has a different speed compared to the universe around us. Not only in the rotation, but also here, the fixed starry sky has a special meaning.

Let's take the normal example again, where $C_0()$ flies away and returns. You can also turn off the acceleration by letting the $C_0()$ accelerate in a loop and then let the earth $B_0()$ pass at the desired cruising speed. For the reversal it is a bit more complex, but also possible. An observer $B_1()$ should rest at the intended reversal point on earth $B_0()$. When $C_0()$ arrives at $B_1()$, it stops its clock, accelerates again in a loop, and restarts its clock when it flies past the observer $B_1()$ again on the way back. Since $B_1()$ is resting to Earth Gemini, its clock will run at the same speed as the Earth clock. The time elapsed on this clock, from the time the spaceman twin arrives until it flies by again after the reversal loop, is subtracted from the time elapsed on $B_0()$'s clock.

One can also eliminate this time difference by looping a spaceman launching $D_0()$ and directing it to fly toward $C_0()$ at the desired return trip speed and so that they meet at $B_1()$. At the moment they meet, the time of clock $C_0(x)$ is transferred to the clock of $D_0()$, resulting in the event $C_0(x) + D_0(x)$. The transmission of the time is, so to speak, the symbolic handing over of a baton. When $D_0()$ arrives at the earth, this clock would then show less time than on the earth clock. To avoid discussions about the Earth's gravitational field, one could also fly three spacecraft outside the solar system.

Now what do I understand as **the paradox**? From the special theory of relativity it follows that all inertial systems should be physically perfectly equivalent. Thus the space traveller twin $C_0()$, which represents K', should be absolutely equivalent to the remaining earth twin $B_0()$, which together with $B_1()$ represents K. From the point of view of the $C_0()$, the clock of the $B_0()$ should also run slower. Also for the motion of $D_0()$ from $B_1()$ to $B_0()$ the physically complete equivalence for $D_0()$ and $B_0()$ applies. Also from the point of view of $D_0()$ the clock $B_0()$ should run slower on the way back.

For me it is paradoxical that on both paths the pairs of observers should be physically perfectly equivalent, but nevertheless the sum of the times of $C_0()$ and $D_0()$ is actually smaller than the time that has passed on the clock of $B_0()$. If an experiment were actually carried out in this way, everyone would actually measure it in exactly the same way. If both ways were actually completely equivalent, then the sum of the times of $C_0()$ and $D_0()$ would also have to equal the time of $B_0()$. Something can't be right here.

The problem also exists for another example of the twin paradox, in which, at the time of reversal, an observer $D_0()$ from Earth is made to follow $C_0()$ with a correspondingly higher speed. Then the clock of $D_0()$ is reversed from the point of view of $C_0()$ and has gone exactly as many seconds slower as it results from the point of view of $C_0()$ after the Lorentz transformations. The sequence of events is actually like this, a real experiment would run exactly like this. But here, too, according to the special theory of relativity, both the way there and the way back should actually be of equal importance, and the measurement result is different here too. That doesn't suit me. Since the result is actually not the same, the motion sequences cannot actually be equivalent.

How does that look when transferred to the rotation situation? Here we have no linear motion, but a circular motion along a ring. The B-system is supposed to represent the non-rotating ring. The C system should represent the rotating ring and rotate with 0.6 c tangential speed in order to get clear values. In principle, this can also be done with the earth's rotation speed or with other rings with a much larger diameter. In this system, the clocks are synchronized according to Universal Time Coordinated UTC. The observer $D_0()$ should sit on a third ring, which rotates with a tangential velocity of 0.8 c. He represents the space-faring twin.

Then there is the event $\mathbf{B_0(0) + C_0(0) + D_0(0)}$, marked with dot ● in **Figure 10a**. The world line of the $D_0()$ is shown as a bold line. Then, after 4 s of the B system, $D_0()$ reaches the $C_1()$ at $B_{3,2}(4)$. Since $D_0()$ moves at 0.8 c from the point of view of the B-system, the Lorentz transformation for 4 s results in a time of 2.4 s, which is displayed on the clock of $D_0(2,4)$. A time of 3.2 s results for $C_1()$ at 0.6 c. There is thus the event $\mathbf{B_{3,2}(4) + C_1(3,2) + D_0(2,4)}$, marked with a circle ○.

Now we only let $D_0()$ move with 0.2 c from the point of view of the B-system, then $D_0()$ is back at $C_0()$ after another 2 s of B-time. From the point of view of the B-system, another 1.96 s elapse on the clock of $D_0()$. So there is the event $\mathbf{B_{3,6}(6) + C_0(4,8) + D_0(4,36)}$, marked with box ■.

In order to avoid the acceleration problems, we could also have let an observer $E_0()$ move with 0.2 c against the B-system, arriving here exactly at the moment of the encounter of $C_0()$ and $D_0()$ and then the time of $D_0()$, in this case $D_0(2,4)$ on his clock. From the point of view of the B system, 1.96 s would then have elapsed until the encounter with $C_0()$. So it would show the time $\mathbf{E_0(4.36)}$.

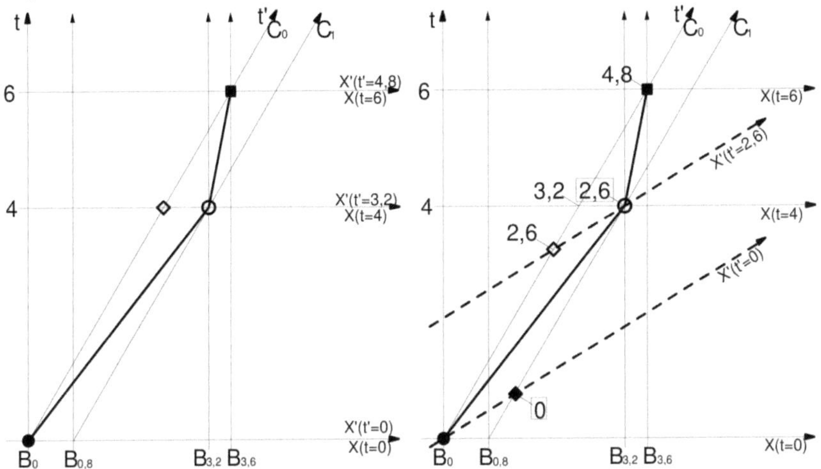


**Figure 10a and 10b:** Twin paradox shown for tangential motion along an equator.

In the C system, the clocks are synchronized as in UTC. From the point of view of the $C_0()$, the motion of the $D_0()$ represents the journey of the astronaut twins. However, he has to realize that the clock of $D_0()$ runs very slowly on the outward journey, and in fact much slower than it appears from the Lorentz transformation would result for him. In the C-system, there are times between the simultaneous events $\mathbf{B_0(0) + C_0(0) + D_0(0)}$ (marked with a dot ●) and $\mathbf{C_1(0)}$ up to the simultaneous events $\mathbf{C_0(3.2)}$ (with a diamond ◇ marked) and $\mathbf{B_{3,2}(4) + C_1(3.2) + D_0(2.4)}$ (marked with circle ◯) 3.2 s, but on the clock of $D_0()$ only 2.4 s.

Between the encounter of $D_0()$ with $C_1()$ and the re-encountering of $C_0()$ and $D_0()$, the event $\mathbf{B_{3,6}(6) + C_0(4.8) + D_0(4.36)}$ (with a box ■ marked), 1.6 s (4.8 - 3.2 = 1.6) elapse on the $C_0()$ clock, while more seconds elapse on the $D_0()$ clock, namely 1.96 s (4.36 - 2.4 = 1.96). On the return journey, $D_0()$'s clock runs faster than the clocks in the C-system. In total, however, only 4.36 s have elapsed on the clock of $D_0()$ according to the Lorentz transformation for the twin paradox and 4.8 s on the clock of $C_0()$.

If we now carry out a spatial concurrency synchronization for the C-system according to Einstein's definition of concurrency, then we only have to put the value displayed on the clock of $C_1()$ back by 0.6 s[20]. This would not affect the actual event sequence of the oscillations of the atomic clocks, or the world lines of any atom, only the actual time values displayed would be different and thus the readings we obtain due to the different spatial synchronization of our measuring instruments. Shown in **Figure 10b**. Then, in the event of $C_1()$ meeting $D_0()$, the clock of $C_1()$ shows only 2.6 s (3.2 - 0.6 = 2.6) and there is the event $B_{3,2}(4) + C_1(2.6) + D_0(2.4)$, (marked with circle **O**). From this spatial simultaneity synchronization of the C-system, between the C-system simultaneous events $B_0(0) + C_0(0) + D_0(0)$ (marked with dot **●**) and $C_1(0)$ would be marked with a rotated box **◆**) up to the events $C_0(2.6)$ (marked with a diamond **◇**) and $B_{3,2}(4) + C_1(2.6) + D_0(2.4)$ (with a circle **O**) which are simultaneous in the C system marked) only 2.6 s had passed and on the clock of $D_0()$ would still be only 2.4 s.

For the return path to event $B_{3,6}(6) + C_0(4.8) + D_0(4.36)$ (marked with box **■**) elapse in the C-system 2.2 s (4.8 − 2.6 = 2.2) and on the clock of $D_0()$ for this route still 1.96 s. This spatial simultaneity synchronization means that the clock of $D_0()$ is also slower for the return route. From the C-system point of view, the clock from $D_0()$ moves faster on the way back from $C_1()$ to $C_0()$ than on the way there, so the time dilation must also be larger, resulting in a larger time difference of 0.24 s for the return trip to 0.2 s for the first trip. With this synchronization, the calculations also agree with the Lorentz transformation from the point of view of the C-system. The twin paradox of the inertial systems under Einstein's definition of simultaneity can therefore also be understood metrologically in the case of rotation. Here, however, we know that the westward-moving clock actually goes faster. Here the paths of the reversing twin are not actually physically equivalent.

There is a simple explanation for the twin paradox if one admits the principle of motion on the gravitational field. Then the twins B and C move towards the gravitational field with the velocity x. If we now

---

20  Comparable to turning the clock back from summer time back to winter time in autumn.

randomly accelerate one of the twins C in the direction of the x-motion, then it not only changes its speed in relation to the other twin, but also in relation to the gravitational field. He's getting faster. The only decisive factor for the result, which clock ran slower when they met again, is which of the two turns back. To turn back, twin C has to expend energy and decelerate again, and in the second phase is slower to the gravitational field than twin B, which was left behind.

After the Lorentz transformations, the time dilation increases exponentially. Thus, the effect of going slower at the higher speed is greater than the effect of going faster at the lower speed. That's why less time has passed on his watch overall.

The Lorentz transformations have their idiosyncrasies. It also doesn't matter whether the relative speed for the outward and return journeys is the same. This only has an effect on the size of the time difference, but not on the trend. Just as in Figure 10, the outward and return journeys are not shown with the same speed.

The observer C could also return with a speed greater than x relative to the gravitational field, then he would be faster than B in both directions, so it is also clear why less time has passed on his clock. C also does not have to accelerate in the direction of the x-motion, the more the angle of the acceleration direction to the x-direction approaches 90°, the more the speed difference for the outward and return journeys equals. At 90° it is the same on both paths, but faster on both paths.

The problem of the twin paradox lies in the spatial simultaneity. When moving around a ring, the spaceman twin can always complete the orbit and compare its clock with the twin left behind before returning. This means that the comparison of the clocks for the outward and return journey can be clearly determined for each observer separately by means of one-place-one-time measurements. However, even the simple travel route is clearly measurable and allows all observers together only three possibilities as a result.

1. From the separation until the reunion there were the same number of cesium oscillations on both clocks.

2. There were fewer cesium oscillations on one watch and thus clearly more on the other.

3. An inverse behaviour of the clocks is possible, but then equally valid for both.

The same goes for the way back.

At no phase of the test procedure are the motions physically equivalent for both observers. There is only one special case: when both observers move in opposite directions to a non-rotating ring with the same speed. Then, when they meet again, the same time has passed for both of them. If both turn around with the same expenditure of energy and then come back with perhaps a different speed, but again the same for both, the motions for both observers in both phases are in fact physically equivalent. But then their clocks also show the same amount of time that has passed.

If you limit yourself to a section, the same applies in principle. It is just not possible to measure unequivocal one-place-one-time events here, because the clocks reverse at different points in space, meaning that a comparison of the clocks is only possible using defined measurement instructions. The comparison is arbitrarily dependent on the concurrency definition, and thus only as valid as the measurement instruction.

Events that are spatially apart can only be causally placed in a chronological order with the help of the fastest available means of information. Within this time loop, from sending to receiving again, a simultaneity cannot be determined in more detail. This applies to any speed, even to the transmission of information faster than light. If one were to discover such a thing, it would also causally limit the possible relative speed for light in straight-line motions. Corresponding to the direct signal comparison at a location when a flash of light has completely circumnavigated a ring during rotation.


# 5. Space, time and space-time

### *5.1 Their starting conditions*

If one starts from the absolute constancy of the speed of light and the equality of all inertial systems, one arrives logically at the Lorentz transformations and at the logical consequence that there can be no speed higher than the speed of light. You have to follow Einstein without contradiction. With information signals that are transported faster than the speed of light, no matter on what physical basis, world lines could also be created, but which would then be incompatible with Einstein's definition of simultaneity for causal reasons. The spatial simultaneity would be more narrowed by faster-than-light signals.

Thus, even for observers moving linearly to one another, the speed of light cannot actually be the same in both directions for everyone.

It must be clarified from which basis one proceeds. Does the different rate of the clocks at different altitudes to the earth, i.e. in different areas of the gravitational field, also mean a change in time itself? Then why does the pendulum clock go slower on the mountain? I assume that the time itself is not affected by this. Only the rate of the watch changes due to the different environmental conditions. This can then be called local time or proper time[21].

Even if a clock turns backwards, the timeline of its world line continues to advance. If someone assumes that their world line will also reverse, they leave the path of causality and anything is possible. Surely this can be represented in linguistic reality, but for me it is not a process that actually occurs in natural reality.

The relative relationships of the observers to each other are undoubtedly correctly described by the Lorentz transformations. The relationships between the observers resulting from the Lorentz transformations are very clearly represented in Minkowski diagrams. Using these diagrams, I had already presented some problems in the previous chapter 4.

All causally incomprehensible confusions arise solely from the postulate of the absolute constancy of the speed of light and are only inherent in the conditions of the SRT.

Whenever gravity comes into play, special relativity does not apply. Where in the universe around us is there an area where there is no gravity? I would like to quote Einstein from his speech given on May 5, 1920 at the Reich University in Leiden:

*"No space and also no part of space without gravitational potentials, because these give it its metric properties, without which it cannot be thought at all. The existence of the gravitational field is directly linked to the existence of space."* (translated by me)

In this I fully agree with Einstein as far as human measurable space is concerned.

If you have a medium, no matter what it consists of, on which all mass particles act and which affects all mass particles, there is a structure for which one can inevitably specify a state of motion. With a medium like water or a gas, this is not a problem because you can observe the individual mass particles or measure the resistance as they move through the medium. We can observe the effect of the

---

21   [11] S.88, [22] S.103

gravitational ether, but that's not enough to determine the speed. To describe it graphically, it would be like trying to determine the speed of the water from the motion of sand and gravel in a river bed, even if you cannot observe the water itself.

How could one detect a motion against the gravitational ether? Not an easy task, and it is also impossible with ideal clocks and light alone, as can be followed from the Lorentz transformations.

Einstein also says in his speech mentioned above:

*"The Maxwell-Lorentz theory of the electromagnetic field served as a model for the space-time theory and kinematics of the special theory of relativity. This theory therefore satisfies the conditions of the special theory of relativity; however, viewed from the latter, it takes on a new appearance. Namely, let K be a coordinate system, relative to which the Lorentz ether is at rest, the Maxwell-Lorentz equations are initially valid in relation to K. According to the special theory of relativity, however, the same equations are also valid in a completely different sense in relation to every new coordinate system K', which is in relation to K in uniform translational motion[10]. The anxious question now arises: Why should I distinguish the system K, to which the systems K' are physically completely equivalent, in theory from the latter by assuming that the ether is at rest relative to it? Such an asymmetry of the theoretical structure, which is not matched by any asymmetry of the system of experience, is intolerable for the theoretician.. It seems to me that the physical equivalence of K and K' with the assumption that the ether is at rest relative to K but is moving relative to K' is not exactly wrong from a logical point of view, but is unacceptable."* (translated by me)

In chapter 5.4 I would like to describe a method with which one can measure the motion against the gravitational field. That would give the experience a difference between K and K'. The measurement then makes no difference between a rotational motion and a linear motion. Both would be motions to the gravitational field. And just as in the case of rotation, in which the clocks moved to the gravitational field run slower, they are then also measured as running slower in linear motion.

## *5.2 What does rotation mean?*

Let's build two ring-shaped scaffolds. In principle, the diameter can be arbitrary. Let's start with the diameter of the earth's orbit around the sun. An observer sits on the rings with his watch. Let's call them B and C. In order not to be confused by any mass effects, let's let the whole thing take place far away from mass accumulations, for example between two distant galaxies. What now decides whether the observers measure a centrifugal force or not? Let's assume they don't measure centrifugal force. Now we accelerate the ring of observer C. Now B and C rotate towards each other. Experience teaches us that C now measures a centrifugal force that increases with increasing speed. Why is C measuring centrifugal force and why isn't B?

If space is really empty and you can't move to anything, then only B moves against C. Now what could be causing centrifugal force in the empty space at all and decide where the centrifugal force occurs.

In general you will find that no centrifugal force is measured if you are not rotating to the surrounding galaxies. The faster you rotate to the galaxies, the greater the centrifugal force.

Empty space is spoken of again and again.[22] How is it defined? With many calculations one pretends that it really is empty. But then again it is said quite unobtrusively that the gravitational field is everywhere in the universe[23].

Max Born [6] writes on p.73: *"...This fact, that certain physical processes are propagated through space, led early on to the hypothesis that space is not empty at all, but is filled with an extremely fine, imponderable substance, the ether, which is the bearer of these phenomena. As far as one still uses this concept of the ether today, one understands nothing other than the empty space affected by certain physical states or fields. If we wanted to commit ourselves to such an abstract formation of concepts right from the start,..."*

In order to be able to assume "certain physical states", something must be present. Whatever that is, and while we can't pinpoint it, it's very concrete, not abstract. We can see it not only in its local but also everywhere in its effect.

Thus there is no empty space in the universe surrounding us, because this is filled with the gravitational field. So there is only one low-mass area in the gravitational field. The gravitational field is then also what determines the centrifugal force.

---

22 [12] p.462 "Free fall in a gravitational field is equivalent to force-free motion in empty space."
   [21] p.24 "Why the inertial mass ... should also occur in a world where there is no gravity"
   p.26 "Only in the total absence of gravity (as is ideally assumed in STR) (translated by me)

23 [12] p.530 "In the entire universe there is no place that would be completely free of gravitational fields."

A quote from Einstein's speech mentioned above: "...*This Machian ether not only determines the behaviour of the inert masses, but is also determined in its state by the inert masses.*

*The Machian thought finds its full development in the ether of the general theory of relativity. According to this theory, the metric properties of the space-time continuum are different in the vicinity of the individual space-time points and are partly determined by the matter existing outside the area under consideration. This spatio-temporal variability in the relationships between scales and clocks, or the realization that "empty space" is neither homogeneous nor isotropic in physical terms, which forces us to describe its state using ten functions, the gravitational potentials $g_{mn}$, probably finally eliminated the notion that space is physically empty. With this, however, the concept of the ether has again come to a clear content, admittedly to a content that is far different from that of the ether of the mechanical undulation theory of light. The ether of the general theory of relativity is a medium which itself is devoid of all mechanical and kinematic properties, but which helps determine the mechanical (and electromagnetic) events.*" (translated by me)

But if the gravitational field is everywhere, then one also moves everywhere towards the gravitational field. Whether between two galaxies, around the sun or directly towards the sun.

Sending sound signals in the atmosphere, far, near, or perpendicular to or away from the earth, to me they're all equivalent, even if the speeds of sound are different. From my point of view, the above-mentioned motions of a body that is not supplied with energy or that does not emit any energy are therefore just as equivalent. I see no difference in the motion relative to the sound transport medium or the motion relative to the gravitational field. Therefore, for me, a speed of light that is constant to the gravitational field does not contradict the equivalence principle.

Let's break down the effects in detail. Every time B and C meet again during their orbits, they can compare their **clocks** directly with each other. Then that clock went slower, which rotated faster to the surrounding galaxies. In this case, this also corresponds to a higher centrifugal force. If both measured the same centrifugal force, then their clocks went at the same speed. For example, if they are not rotating or if they are rotating in the opposite direction at the same speed. In general, it can be said that they never have equal rights to move, except in the last-mentioned special cases.

What about **light signals**? We imagine a mirror wall around the rings of the observers, along which they can send their light signals in both directions. When they meet, B and C emit a flash of light. Since the flash of light needs some time for the orbit and the observers keep moving during this time, it can be stated that the part of the flash that moves in the C-direction (I call it c-flash) reaches B first and then C and the part moving in the opposite direction (I call it b-flash) reaches C first and then B. Since light propagates with a uniform front, it is clear that the flashes can only reach one of the two observers at the same time after a complete orbit. From actual observations we can conclude that this is the observer who is not rotating to the surrounding universe. For the observer who is moving, the speed of light is not the same in both directions.

All of these logically causally developed conditions are actually confirmed in terms of measurements by experience with satellite navigation. Here there is clearly a preferred rest system for the rotation, which also applies to the motion of the light. If the observers' clocks are synchronized with each other according to Universal Time Coordinated, a light signal sent from a western observer to an eastern observer takes longer for this direction than in the opposite direction.

How can the rotational motion be determined at all?

1. by centrifugal force.

2. by comparing the arrival of light signals that had been sent in both directions one circle around the circular path. This corresponds to the Sagnac effect or the Michelson-Gale experiment.

3. by observing astronomical events and relating them in time to the angle at which one sees them from one's position.

Let's transfer this to the rings and first let the two rings rest side by side against the surrounding events in the sky. Let's only look at a **small, equally sized section of these rings**, at the ends of which observers sit with their watches. Each of the observers has an ideal atomic clock, all of which are calibrated in the same way, i.e. they show the seconds according to the definition. In addition, we are building two spaceships of the same length as the sections, in each of which there are two observers at exactly the same distance. All four pairs of observers carry out a spatial synchronization of their clocks according to Einstein's definition of simultaneity.

Then one of the two rings begins to rotate. And one of the spaceships is accelerating in a loop so that it is moving tangentially past the stationary ring with the spaceship next to it at the same tangential velocity of the rotating ring. We call the non-rotating ring K, the rotating ring K', the rocket resting on K R and the rocket flying by R', see **Figure 11**. After the acceleration, the observers K' and R' realize that they at the current synchronization do not measure the light in both directions as equally fast. In order to achieve this, the observers K' and R' must spatially synchronize their clocks again.

According to **Einstein's definition of simultaneity** (see also Chapter 2.3), a third observer would now have to measure the middle between the pairs of observers and observe here which events from the observers reach him at the same time. Based on the constancy of the speed of light, it is equivalent to sending out a flash of light in the middle of this observer. The arrival of the flash of light at the other observers would then be simultaneous. A simpler method to do this, especially if additional observers are integrated or attached to an already existing chain, is the following:

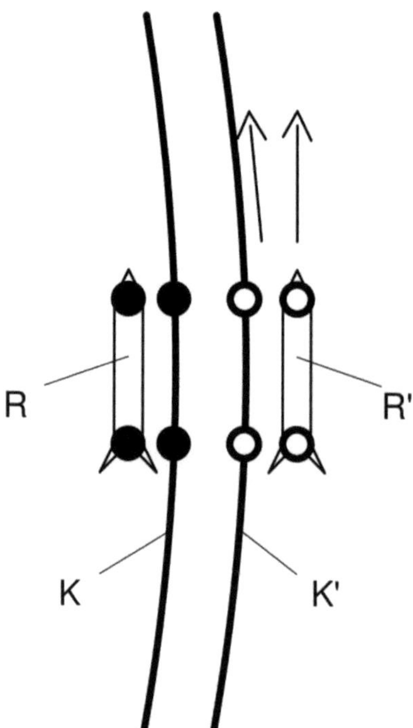


**Figure 11:** K and K' represent parts of the ring constructions. R and R' are adjacent spaceships. R rests on K. R' moves tangentially past K and R with the tangential velocity of K'. K' moves past the observers on K and R on the circular path. Why should the spatial simultaneity in R' be different than in R, K and K', where the spatial simultaneity is the same for causal reasons?

To do this, the first observer sends out a time signal, which is reflected by the other observer, and the first observer measures the time it took the signal to travel there and back. The second observer now sets his watch so that when the time signal arrives it shows the time of the time signal plus half the time for the outward and return journey. Half of the time for the outward and return journey also corresponds to the distance between the two measured in light seconds (Ls).

If all clocks were spatially synchronized with each other before the start, then the clocks of K' and R' must be resynchronized. In order to achieve synchronization according to Einstein's definition of simultaneity, the front clock of K' and R' must be put back by a certain amount or the rear clock must be advanced by the amount. The amount depends only on the distance between the clocks and the extent of the changed speed.

If the observer groups now move past each other, arise for everyone what Einstein called "simultaneous events". For example, when a train passes the station and an observer on the train or on the platform reads his watch and the station clock at the same time. Or as I call them, one-place-one-time events. They are encounters between two or more observers at one time and one place, even if they belong to different observation systems that move against each other. The observers read the time displayed on their watches for the encounter. All observers, including those farther away, must all reconcile this event with their own measurements. This applies to the pairs of observers on the rings as well as in the spaceships.

Since the clocks on the rings and in the spaceships were synchronized according to Einstein's definition of simultaneity, all pairs of observers measure the distance between the pairs of observers who are moving towards them as shortened and the clocks of the others as running slower. This also applies to the rotation[24]. Under this spatial synchronization according to Einstein's definition of simultaneity, the rotating observer pair also measures the distance of the stationary pair as shortened, which, however as not moving, actually has a greater distance.

You can fill in the complete rings with the same distances, e.g. standard meter. After accelerating for rotation, you can see that there are gaps between the standard meters or a large gap in the ring.

---

24  In the case of rotation, Einstein's definition of simultaneity is not consistent with actual observations, which is why the earth clocks in Universal Time Coordinated UTC or satellite navigation are not synchronized accordingly.

That is the fundamental problem when measuring instruments can only be calibrated and synchronized according to defined instructions and then the measured value and the actual being differ. This only ever attracts attention when measured values are obtained on different physical basis in one and the same test and at least one of the measurements does not depend on the defined instruction. Whereby there are the 3 possibilities mentioned above for the rotation, as well as the length contraction mentioned here. In the case of linear motion, however, none of the effects can be measured as an indication of motion.

The observer pairs on the non-rotating ring and in the relatively stationary spaceship measure the distance of the other observer pair as equal in each case. With the rotating ring and the spaceship moving at relatively the same speed, there are lateral motions that should not be underestimated given the required precision, but which only play a subordinate role for spatial synchronization. Since also these pairs are spatially synchronized in the same way, they also measure the distance of the other pair as equal.

We know that the observer group on the rotating ring is actually contracting in length and the clocks are actually slowing down. Nevertheless, the rotating group of observers measures the resting group of observers from their newly synchronized simultaneity level after the acceleration, also as shortened. Likewise the non-rotating clocks on the short stretch between them as going slower.

When the observer groups complete the circumnavigation and meet again, the non-rotating group finds that the clocks of the other group show exactly the less time that has resulted from the extrapolation of the first measurement. The moving group of observers again measured the clocks of the others on the short stretch between them as rating more slowly. However, it has to realize that a lot more time (displayed seconds) has passed on these clocks in the meantime, so they must actually have gone faster. That means something must be wrong with their measurements.

In the case of the spaceship moving tangentially to it, however, it should only be an apparent contraction, because one has no other possibility to determine the spatial simultaneity with the straight-line motion. Therefore, the spaceship R' shown in Fig. 0 as an inertial system should be on an equal footing with the spaceship designated as R?

If the observers sitting on the rings confine themselves to one section and only make measurements with ideal clocks and light signals within this section, then they cannot decide which pair of the two is moving and which is stationary. This is inherent in the idiosyncrasies of the Lorentz transformations.

In the case of a uniform translational motion, the three possibilities for determining the rotation motion mentioned above, do not occur. They also don't meet again without one of them changing their state of motion, so they can't compare their clocks with each other again. But why should the pair of rockets that moved with the rotating ring pair not actually but only appear to be shortened?

Due to the observations that can be made during the rotation, this pair of observers must recognize that spatial simultaneity is not causally given even on a section according to Einstein's definition of simultaneity. The clocks in Universal Time Coordinated (UTC) are therefore not spatially synchronized like that. In synchronization with UTC, a light signal in a westerly direction between two clocks takes less time than in an easterly direction.

In the quote mentioned in Chapter 4.1, Einstein says that there is *"no asymmetry in the system of experiences"* when comparing the systems K with the systems K'. In the example given here, there is an indication of asymmetry. The faster the ring rotates, the greater the actual length contraction and the slower the clocks go. The microwave background radiation is measured with an increasing blue shift forwards and with an increasing red shift backwards. The light sources revolve around the observer according to the rotation. In the case of the uniform translational motion, this does not rotate, but there is also in the same amount a blue and red shift in the microwave background radiation. If we have a ring diameter of 320,000 light years, then the circular motion needs about 1000 years even at 90% of the speed of light for 1°. This is hardly noticeable during a measurement phase of a few minutes. However, the blue and red shift of the microwave background radiation has the same measured value for both observers.

If the pairs of ring observers and spaceships at rest measure themselves as being of equal length and the rotating pair of ring observers actually contract in length and both observe the microwave background radiation as equally asymmetrical at the moment of tangential contact, then one should be able to assume that it is also the case with the spaceship in uniform translational motion is also an actual length contraction. Even if that doesn't immediately mean that the equivalence of K and K' is unacceptable, it should be a reason to think again about whether the previous views can remain so unchallenged.

## 5.3 Rotation and gravitational field

Here I would like to present my model. Contrary to what Einstein assumed, it should be a medium to which the concept of motion can be applied. What it could consist of, I would like to leave open. Figuratively speaking, I want to pop black powder without knowing what it's made of. Or I would like to launch my sailing ship without knowing what this medium water or air consists of and still sail with it.

Einstein described something similar to what I described with the 2 rings in the book [10] p.52:

*"We are again assuming very special cases that are often used. There is a space-time region in which there is no gravitational field relative to a reference body K with a suitably chosen state of motion; with respect to the area considered, K is then a Galilean reference body, and the results of the special theory of relativity apply relative to K. We think of the same area related to a second reference body K', which rotates relative to K uniformly. In order to fix the idea, let us imagine K' in the form of a flat circular disc which rotates uniformly around its centre in its plane. An observer sitting eccentrically on the circular disk K' feels a force that acts outwards in a radial direction, and which is interpreted by an observer who is at rest relative to the original reference body K as an inertial effect (centrifugal force). However, the observer sitting on the disk may perceive his disk as a "resting" reference body; he is entitled to do so on the basis of the general principle of relativity. He interprets the force acting on him and generally on bodies that are at rest relative to the disk as the effect of a gravitational field. However, the spatial distribution of this gravitational field is such that it would not be possible according to Newton's theory of gravitation. ..."* (translated by me)

There should be no gravitational field for K, but one for K' that is quite extraordinary? Why shouldn't K also feel this gravitational field? Einstein had also said (in more detail p.7) *"... According to the general theory of relativity, space without ether is unthinkable;..."* Does everyone who considers themselves at rest have their own ether, or rather gravitational field? Where does it even come from? Looked at another way, we omit K, why should K', who sees his disc as a "resting" reference body, measure a centrifugal force? K should also have the same circular disc and we add a K" that should rotate in the opposite direction to K' with the same speed. They all put mirrors on the edges of their panes. They arrange it so that they always meet at the same point. When they meet, they emit a bolt of lightning. Then they find out that this always arrives at K again from both sides at the same time.

If K' considers himself at rest, he must now reconcile the strange motions of K and K" with his gravitational field and the absolute constancy of the speed of light. If we do that in the universe that really surrounds us, it also has to reconcile this with the galaxies rotating around it, which also reach speeds faster than light on the horizon.

If K and K' meet again at each orbit, there are only three possibilities for both of them as to how their clocks went. The clocks went at the same speed, K"s clock went faster or K's clock went faster. From K's point of view, K"s clock must slow down. This means that K's clock is also faster from K"s point of view. Mathematically, it may still work out that the faster motion of K's clock can be calculated from the point of view of K', which regards itself as being at rest.

From K's point of view, the clocks of K' and K" must be running at the same speed. This means that the observers K' and K" must also determine that their clocks are running at the same speed. How does K' explain the faster motion of K's clock and the equally fast motion of clock K", which is moving almost twice as fast.

In contrast to the inertial motion of the special theory of relativity, K and K' do not have equal rights here. There is a preferred state of rest for all rotating disks. Just as it is also determined in satellite navigation.

The principle of relativity that Einstein mentions here for the rotation cannot be that of the special theory of relativity, see p.21. "*Only the motions of the bodies relative to other bodies can be determined, but not the motions of the bodies relative to a **preferred reference system**.*" Here K clearly represents a preferred reference system. The observers rotating at the edge of the disk of K' can see their Accurately determine motion speed versus K. They don't even need K. By determining the centrifugal force, they can calculate how fast they are moving.

Einstein introduces the whole thing with a "*space-time region in which no gravitational field exists relative to a reference body K with a suitably chosen state of motion*". So does it really not exist? Or does it just not lead to any centrifugal force because the observer does not rotate?

In the reality of the universe surrounding us, it is certain that this state of motion cannot be chosen freely, but is determined by the surrounding fixed starry sky.

As Einstein said, there must be something between the mass particles that they act on, but that also acts back on the mass particles. A gravitational field in the sense of E. Mach. Let's assume, contrary to Einstein, that there can be actual motion to this gravitational field. Then we can also explain the rotation with the gravitational field. A system that does not rotate to the gravitational field does not measure any centrifugal force. The faster it rotates, the greater the measurable centrifugal force. Let's take the ring system mentioned in the previous chapter again. Only one state of motion can be the state not rotating to the gravitational field. All rings rotating for this must also measure a centrifugal force.

The centre of the rings could also move towards the gravitational field. Regardless of the speed at which a non-rotating ring moves towards the gravitational field, an additional rotation, with the serpentine line that then occurs, always leads to a higher speed towards the gravitational field. Thus, the motion to the gravitational field is the decisive factor as to how far there is a centrifugal force, an actual length contraction, or a clock actually running slower. Corresponding is also valid for the uniform translational motion of the rockets. The only problem here is that we have not been able to measure this straight-line part of the motion so far. In the following chapter 5.4 I would like to present such a possibility.

What the gravitational field actually consists of I cannot say. Let's assume it would be composed of all fields of the mass particles. Then this field also exists in the whole universe, as well as in the middle between distant galaxies. Then the motion of mass particles to this total field must also have an effect on it. The sum of the gravitational field formed by the surrounding universe and the gravitational field of the moving mass could result in a gravitational field that moves slightly towards the surrounding universe.

Einstein interpreted the perihelion rotation of Mercury as a superimposition of the gravitational field. I see it as an actual turbulence of the prevailing gravitational field, roughly analogous to a water or air turbulence. Then rotating masses would also lead to a rotation of the gravitational field. A ring not rotating in this field, in which no centrifugal force is measured, would then rotate in relation to the area outside of this gravitational area. Thus, in a rotating mass system, the centrifugal force would be much lower than would be expected for the speed measured from the outside. A galaxy where the masses all rotate in one direction, such as a **spiral galaxy**, for example, would rotate faster than its mass in relation to the surrounding universe. So no dark matter is needed to keep the galaxy from flying apart. In galaxies in which the stars do not move in uniform mass fluxes, but rather irregularly, their speed of motion should also correspond more closely to the observable mass of the galaxy.

It is questionable whether new formulas have to be found for this description. The general theory of relativity already describes a measurable Schiff effect for the rotation of the earth in the gravitational field determined by the sun and the Milky Way[15]. In my view, this would not only be a superposition of a theoretically rigid gravitational field, but an actual rotation of a time-trackable medium that forms the gravitational field.

Here the gravitational field is essentially formed by the sun and the Milky Way and is only influenced more strongly by the rotating shell of the earth and hardly influenced by the core. If the masses that determine the local gravitational field are all moving in the same direction, as in spiral galaxies, this effect should be complementary and larger.

A similar effect would be the perihelion rotation of Mercury. Here, too, the gravitational field would rotate locally with the sun and produce this effect.

Another phenomenon can be explained with it. If the stars in a spiral galaxy move freely in the gravitational field present in the region (even if dark matter were present), the stars should have completely mixed with each other after five rotations and the spiral arms should no longer be recognizable. However, it seems that the galaxies are developing more towards the spiral galaxies. This may be because the gravitational field is entrained by large clusters of stars, favouring the motion of other stars in that direction. Stars that have lost touch with this group fall back and are collected by the next arm. This would also increase the stability within the arms so that they do not break apart. The effect of gravitational field entrainment may be small, but even the Coriolis force is weak, yet pervasive in weather and riverbank erosion.

Another effect could be explained: the fly-by anomaly:

A satellite experiences a deflection in passing a planet, like comets in orbiting the sun. As the planet moves in the solar system, the satellite experiences additional acceleration in the direction of the planet's motion. However, the space probes are still a little faster after the passage than the standard calculations result in, the so-called fly-by anomaly.

The Schiff effect, which results from the standard calculations, has already been clearly demonstrated. This rotation effect is mainly caused by masses moving very slowly at equator. I assume that the gravitational field actually twists slightly in shape and this creates the ship effect.

In the direction of the motion of the earth around the sun, however, the entire mass of the earth takes the field with it at a much higher speed. This means that the locally prevailing gravitational field would move a little in relation to the surrounding gravitational field in the earth's motion direction. A body would receive an additional acceleration in the motion direction of the earth due to the entrainment of the gravitational field, which naturally weakens with increasing distance. This should be independent of whether it passes in front of or behind the earth. In any case, the additional acceleration should be in the motion direction of the earth.

The Schiff effect, Mercury's perihelion, the fly-by anomaly, and the over-rotating galaxies could all be based on the same physical principle. Since the Schiff effect and the perihelion rotation of Mercury are already described by the general theory of relativity, the question is whether new formulas are needed for the other problems, or whether the existing ones are enough, they just have to be applied in a slightly different way. In the case of the fly-by anomaly, one should not only look at the earth and the satellite in isolation. One would have to regard the earth as a body circling on the earth's orbit, which thus produces an "earth's-orbit-ship-effect", which then causes the fly-by anomaly. With the galaxies a much more difficult task. Here one would have to consider all stars individually and in total as masses rotating around the centre of the galaxy. In order to determine the rotation speed of the gravitational field of the galaxy, you certainly don't need new formulas, you just have to have the courage to set the value for dark matter to zero and the result is the rotation speed of the gravitational field. It would be interesting to see if one could bring all 4 phenomena together with the existing formulas.


### 5.4 Satellite navigation and the gravitational field

If, according to Einstein's statement above, the gravitational field or "this ether (...) were not endowed with the characteristic property of ponderable media (...) to consist of parts traceable through time", then of course there would be no motion against this field. However, should it be the case that a state of rest or a motion against this field is possible, then this must also be able to be determined.

The relative conditions in the universe surrounding us are undoubtedly described by the Lorentz transformations. However, the consequences that result from this cannot be grasped with our feelings or everyday experiences. Even technically competent physicists have their difficulties.

In the book by R. Sexl and H. K. Schmidt[24] on p.27 it is framed: *"The synchronization of the clock network of the world confirms the principle of the constancy of the speed of light. According to the ether theory, constantly changing runtimes of the time signals would be expected, which are not observed experimentally."* (translated by me)

Exactly such an effect (changing runtimes) would be incompatible with the geometry of the Lorentz transformations. Let's imagine all inertial systems being equal. Then I call one of the inertial systems ether. That would not change anything in the mathematical geometry. Now let's imagine we could measure the motion to the gravitational field. Then I pick out the inertial frame that is at rest to the gravitational field and call it ether. That would not change anything in the mathematical geometry of the Lorentz transformations.

During the experimental observations and their mathematical processing, there is no change in the propagation times for time signals. Nothing would change if one could determine that one of the inertial systems is at rest with respect to the gravitational field. Perhaps this will make it easier for theoretical physicists to follow this chapter mentally.

As I wrote above, the Sagnac effect, which must undoubtedly exist for the earth's orbit around the sun, cannot be determined even by such a precise instrument as the Global Positioning System. Here only the Sagnac effect within the rotation of the earth system can be determined. This is not because the effect is covered by any correction value or because the measurement accuracy is not sufficient, but because of the characteristics of the real conditions as described by the Lorentz transformations.

Only ideal clocks and light-speed signals are used in the satellite navigation. Even if the distance between the clocks changes constantly due to the orbiting of the satellites and one-way measurements are also made, the basic conditions correspond to the Michelson-Morley experiment, the one-armed Kennedy-Thorndike experiment, or a non-accelerated one-arm one-way experiment.

In addition, the effect of clock transport within satellite navigation must be taken into account. As Einstein already said, the slow clock transport corresponds to his definition of simultaneity. The rate of a watch thus changes during transport in such a way that it remains in the spatial synchronization level according to Einstein's definition of simultaneity. As a result, their displayed time during transport on earth migrates out of the spatial synchronization of the clocks in UTC. However, the overall system of satellite navigation moves in a spatial simultaneity level corresponding to Einstein's definition of simultaneity. This means that the satellite clocks also move in this level, only their relativistic speed has to be taken into account, since this is not a measurably "slow transport".

All measured values are only obtained within the system. The same would apply to a train travelling around the earth along the equator. Even if we know that the west-moving clock actually runs faster and that the light actually takes longer from west to east than from east to west when the train is standing, we could not determine this with measurements made solely within the train.

It doesn't matter whether the concept of motion applies to the ether/gravitational field, which Lorentz started from when developing the Lorentz transformations, or not, which Einstein started from. The relative conditions existing in the natural reality of the universe surrounding us are described by the Lorentz transformations, otherwise the negative outcome of the Michelson-Morley experiment (MME) and all its derivatives could not be explained.

Mathematical formulas are difficult to understand. It is even harder to see the consequences of these. Therefore, based on a group of observers sitting on a non-rotating ring, clear events are to be calculated here. What clear events are has already been described in chapter 2.5. Every observer who is moved to do so must then also accommodate this in his measurements without contradictions arising.

Let's come back to the experiment of two rings circling parallel next to each other, far away from larger masses. No centrifugal force is measured at the edge of one ring. We want to call this state non-circling.

Let's transfer the insights we gained from satellite navigation and Universal Time Coordinated UTC to this experiment. To measure time, we use atomic clocks that correspond to light clocks. We don't want to measure spatial distances at first.

Let's position the atomic clock A at the observer A on the non-rotating ring A. On the ring B, which rotates to the east in the sense of the Earth's rotation, the atomic clock B at the observer B.

The atomic clocks should show the time according to the defined second. Then clock B shows less elapsed time each time it encounters A. On all clocks positioned on rotating rings, the seconds pass more slowly. The faster they circle, the slower the clocks go. The encounters of the clocks are definite events. It follows that for observer B, too, his clock B runs slower than clock A.

Now, when A and B meet (Fig. 12 - 1), they emit a flash of light that is supposed to propagate along the rings in both directions. Then only in the case of the non-circulating ring A do the lightning components moving in opposite directions reach the observer A from both directions at the same time (Fig. 12 - 3). Light spreads with a uniform front. Nothing else has been observed in the universe so far. Then the lightning components cannot reach B, which is moving towards A, from both directions at the same time. The part of the lightning propagating in west direction, i.e. contrary to the motion direction of B, reaches him first (Fig. 12 - 2) and then the part of the lightning propagating in east direction (Fig. 12 - 4). This also corresponds to the findings from satellite navigation.

The observers should reflect the lightning components when they are reached by them. A thereby simultaneously produces the reflections A(**ew**) and A(**we**) for all observers (Fig. 12 - 3). These reach for all observers A again at the same time (Fig. 12 - 5).

B first produces the reflections B(**we**) (Fig. 12 - 2) from the lightning component that initially propagated in a **w**esterly direction and after the reflection moves in an **e**asterly direction. Only a little later does the part of the lightning that has moved in an **e**asterly direction reach it. With this he generates the reflection B(**ew**) (Fig. 12 - 4) which then moves in a **w**esterly direction. These reflections reach B simultaneously (Fig. 12 - 6), no matter how fast he orbits. He must not change his speed during the course of the experiment.

Every other observer in the universe must observe these unequivocal events in the same way. Nobody reads different times on the clocks than the observers themselves. And the arrival of the lightning takes place at no other place, relative to the surroundings of the respective observer, than for the observer himself.

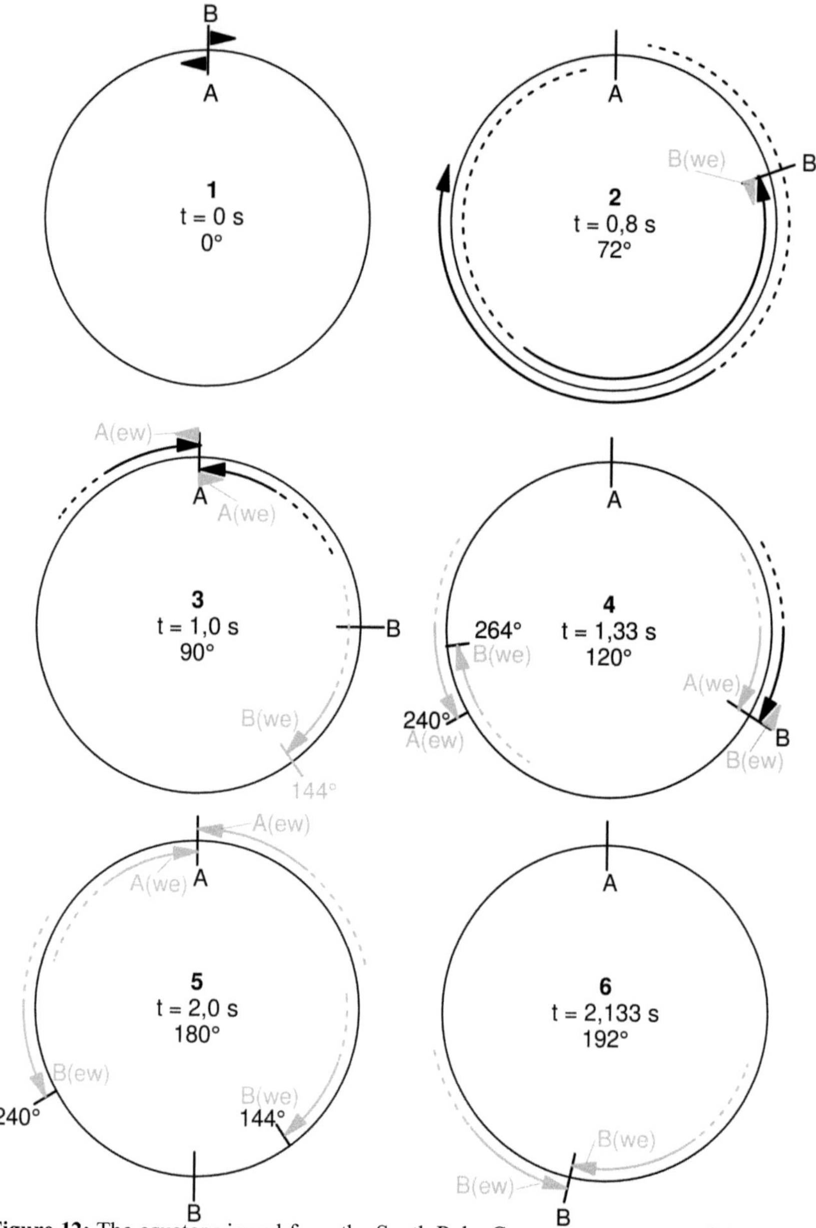


**Figure 12:** The equator viewed from the South Pole. C moves at a quarter of the speed of light.

To do this, the first observer sends out a time signal, which is reflected by the other observer, and the first observer measures the time it took the signal to travel there and back. The second observer now sets his watch so that when the time signal arrives it shows the time of the time signal plus half the time for the outward and return journey. Half of the time for the outward and return journey also corresponds to the distance between the two measured in light seconds (Ls).

If all clocks were spatially synchronized with each other before the start, then the clocks of K' and R' must be resynchronized. In order to achieve synchronization according to Einstein's definition of simultaneity, the front clock of K' and R' must be put back by a certain amount or the rear clock must be advanced by the amount. The amount depends only on the distance between the clocks and the extent of the changed speed.

If we had already done this when A and B met, then at that time the clocks would read A = 0, BE = 0, but BT = - 0.25820. So BT emits the time signal - 0.25820. This is achieved by BE in Fig 12 - 2. It takes $t = 0.8$ s, i.e. $t' = 0.77460$ s, until then, i.e. when the clock shows BE 0.77460 s. BT would emit the time signal 0 0.25820 s later, so the signal would also reach BE 0.25820 s later, so if the clock shows BE $t' = 0.77460$ s $+ 0.25820$ s $= 1.03280$ s, what corresponds to exactly half for the outward and return journey. In this spatial synchronization, the clocks would also measure the speed of light in the moving ring in both directions at the same speed, even if B can see through the foil and establishes that the lightning bolts do not actually reach B from both directions at the same time.

In chapter 5.6 the whole thing is explained more generally and then with formulas.

No matter how we shield these rings, as long as we don't bring other masses into play, the outcome doesn't change. Mathematics cannot help us to decide in which ring no centrifugal force is measured, the clocks go fastest and in which ring the lightning strikes the observer from both directions at the same time. Only an actual measurement can help us further. If we lift the curtain in this case, we find that all 3 phenomena are associated with a non-circling in relation to the universe that surrounds us.

The relative relationships resulting from the Lorentz transformations are quite confusing. However, since we can assign the measurements to unambiguous events during the rotation, they are unambiguous and equally valid for all observers.

Let's start with atomic clocks that are set up along the equator and are spatially synchronized with each other according to UTC. From this synchronization, the time taken by a light signal from a west clock W to an east clock O is measured as greater than in the opposite direction.

A clock C moving eastwards with respect to the stationary clocks O and W can be regarded as being on a faster rotating ring; one moving westward than being on a slower orbiting ring. So a westbound clock must go faster, and that's what a westbound clock on Earth does.

To leave out all the irregularities of Earth's gravity and atmosphere, let's just imagine the rings again. Starting from the non-orbiting ring, we take all the measurements and use it to obtain a set of unique events, how the clocks on the other rings behave and how the light signals propagate along the rings. All these unambiguous measurement results must then also correspond to the measurements of the observers on the circling rings. No calculations are intended to confuse the reader here, only the relative relationships that result from the Lorentz transformations are presented.

Let us now only look at the section between the clocks W and O. They are now to be spatially synchronized according to Einstein's definition of simultaneity. Since they were previously synchronous in UTC, the easterly clock O, for example, only needs to be put back by the amount n. Then they again measure the speed of light equally fast in both directions. The clock O would then lag behind UTC by the amount n. If we now move a clock C from the western clock W to the clock O, or vice versa, the clock C remains exactly in the spatial simultaneity of these clocks. This corresponds to Einstein's statement that the slow clock transport corresponds to his definition of spatial simultaneity. The displayed value of the clock C at the start compared with the clock W or O differs when it arrives at the other clock O or W only by the amount that results from the Lorentz transformations for this relative motion of C compared to A and B. This can therefore be neglected in the case of slow watch transport. Clock C is measured as going at the same speed as clocks W and O on its way between clocks W and O. This measurement result is only due to the different setting of the measuring instrument for the spatial simultaneity. In fact, it is much slower going in east direction than the Lorentz transformations suggest for this relative motion between clock C versus clocks W and E, and going faster in west direction.

The clock C could move at a constant speed in an easterly direction from W to E and further along the rest of the ring to W again. When it arrived again at W, it would then have rated slower by the amount m. On the further way from W to E it doesn't seem to go any slower compared to W and O, because even when it arrives at O the displayed value of the clocks of C and O differs by exactly the amount m. The clock C could also move in westerly direction at constant speed from clock E to clock W and the rest of the ring back to clocks E and W. Then, when it arrived at clock O, it would have gone the amount m faster than the latter, and when it arrived at clock W, the display differs by exactly this amount. It doesn't matter which part of the circumference of the ring is the distance between W and O. The amount of m depends solely on the circumference of the ring and the tangential speed with which it moves.

Let's increase the distance between W and E so that it spans the entire equator. Then W and O are in the same place. Then W and O could equally determine for all observers that they would have to set different times for spatial synchronization according to Einstein's definition of simultaneity at one location. The difference is exactly m. The displayed times of W and E are only the same in the non-circulating ring. Even with a non-rotating planet, if the clocks were synchronized to UTC, the time signals at E and W would arrive with the same time difference, i.e. the same time would be measured for both directions.

The clocks W and O are now to be mounted on the circling ring B on a beam which can be rotated at its centre. So that the paths of the clocks are the same when turned relative to the ring, the bar must be turned in the tangential plane, which corresponds to turning horizontally on the earth's surface at the equator. First, the clocks should be in sync with UTC. What effects would be observed if the bar were rotated by 180°?

1. The westbound clock O is now ahead of UTC by the value n.

2. The eastbound clock W is now lagging behind UTC by the value n.

3. The clocks measure a longer time for the light signals from W to E than from E to W before turning over.

4. There is no shift in the arrival of the time signals during the turning.

5. After the turning, the clocks W and O continue to measure a longer time for the signals from W to E than from E to W. So the opposite of what it actually is. Shown more clearly in the text below with values for the unique events.

6. Let's perform a spatial synchronization of clocks W and O according to Einstein's definition of simultaneity. For this we have to set back the clock O by the value n. It thus lags behind UTC by the value n. But now the clocks measure the duration of the time signals equally in both directions.

7. After being turned, clocks W and O continue to measure the time duration for the time signals equally in both directions. Clock O is now in sync with UTC again, but clock W lags behind UTC by the amount n.

Of course, the statements do not only apply to rings with the size of the equator. The rings could also match the Earth's orbit around the Sun or be five times the diameter of the Milky Way. Can easily be accommodated in a void without larger masses disturbing. If one considers only a small section of the rings with the latter ring size, the circular motion would hardly be distinguishable from an inertial motion.

In the book by Sexl / Schmidt[24] it says on page 54: "... *We were allowed to consider the earth approximately as an inertial system.*" In the book by Hetznecker[12] it says on page 72: "... *So can we actually call a laboratory on Earth an inertial system under these circumstances?... The earth's surface is an inertial system, and that's a good thing.*" Now you can reverse this equation and say: In sufficiently small sections, inertial systems correspond to a rotation. With rotation, however, one can clearly determine who is rotating and who is not moving tangentially to the surrounding universe. As I have already described in chapter 3.11, this is an impermissible equation. If not in one direction, then not in the other either.

Let's imagine a pair of rings the size of the Earth's orbit around the Sun. The atomic clocks would also run fastest on a ring that did not revolve in relation to the surrounding universe. Whether or not the sun is at the centre would have an effect on the centrifugal force and gravitational influence on clock rate, but not on the motion effects.

Inevitably, the clocks that move faster here at night due to the rotation of the earth rate slower and faster on the day side. The same applies to satellites. The clocks that move faster in the direction of motion when circling around the earth must rate slower and faster in the opposite direction.

As with the bar that is rotated at the equator, this has no effect on the arrival of the time signals. For this reason, this effect cannot be determined within satellite navigation either. In order to determine this effect, one would actually have to construct a ring corresponding to the earth's orbit and synchronize the clocks in it spatially without contradiction. That would correspond to the principle of UTC, just for the earth's orbit around the sun. At this level of spatial simultaneity, the clocks on Earth would also run slower at night and faster during the day. Just as an eastbound clock runs slower and a westbound clock runs faster.

The real implementation of the Michelson-Morley experiment MME happens while it rotates on earth, orbiting around the sun and with the sun around the centre of the Milky Way. It doesn't move in an inertial system. Since the clocks actually slow down during the rotation for all observers and the speed of light on the earth's surface is actually not measured at the same speed in both directions, there must also be an actual length contraction[25], otherwise the MME would not be negative.

Despite all these actual effects, the MME must be negative if the relative ratios correspond to the Lorentz transformations. From the Lorentz transformations it follows that even if we put the MME on rails and accelerate it in an easterly direction with respect to the earth's rotation, or slow it down in a westerly direction, this would have no effect on the negative output.

If we want to achieve a measurable effect, then we have to reduce the MME to a one-way experiment and place lasers or clocks that work equally constantly at the ends of the arms. Quite possible given the precision of today's atomic clocks. Then a clock pair W and E in west-east direction would correspond to one arm of the MME and another clock pair N and S in north-south direction would correspond to the other arm. As described above, spinning such an attempt has no effect. But if we accelerate this experiment, e.g. in an easterly direction, the time signals arrive later and later at clock O and earlier and earlier at clock W. Accelerating in the opposite direction has exactly the opposite effect. It is therefore not possible to tell from this whether the pair of clocks is circling faster or slower after accelerating.

In the case of the clock pair N and S lying transversely to the direction of acceleration, there is no effect during the acceleration.

Let's let several pairs of clocks W and O circle past each other at different speeds. According to Einstein's definition of simultaneity, the pair of clocks on a ring should be spatially synchronized with each other. Then the measurement results obtained during the passage with

---

25 [11] p.251: "But it is already clear that not only the clock rate is influenced by the acceleration or by a gravitational field, but also the length measurement."

mutual observation also correspond exactly to the predictions of the special theory of relativity during the rotation.

The twin paradox is also resolved quite logically. From the spatial synchronization of the pairs of clocks according to Einstein's definition of simultaneity, each clock that moves past them is measured as running slower according to the Lorentz transformations. If such a pair of clocks turns back again with the expenditure of energy, then it must, as described above, e.g. set the front clock back by the value x so that the speed of light is again measured to be the same in both directions. From the spatial synchronization according to Einstein's definition of simultaneity, a clock passing you westwards is then also measured as going slower, although it is actually going faster.

## 5.5 Time - and what do clocks have to do with it?

We take light clocks and atomic clocks, which behave in the same way, as a basis. Depending on their position in relation to the gravitational field, they then move at different speeds. The second they measure is shorter on the mountain than in the valley. They rate faster up the mountain than down in the valley.

Pendulum clocks have the opposite behaviour. They are slower on the mountain than in the valley. The clocks are based on a different physical basis. For centuries, pendulum clocks could be used to tell the time. Atomic clocks are more accurate, but do they have more to do with time than pendulum clocks?

A little excursion into another area to clarify the question. I want to determine the mass of a body. To do this, I measure the body with a beam scale and with a spring scale. I do that again on the moon and get different measurement results with the spring balance. Now I also ask here: What do the measuring instruments have to do with the mass?

I don't want to get too deep into this excursion. Ultimately, different measurement principles (leverage to material strain) are affected differently by environmental changes.

I hope that the willing reader will continue to follow me. I think the same goes for any timing instrument. Pendulum clocks show the time just like light clocks.

When measuring the masses with the beam balance, the mass is, so to speak, measured against itself. With light clocks, the motion of light is measured with itself. As a result, light clocks automatically measure the time as given in the geometry of the Lorentz transformations without having to be adjusted.

Light clocks indicate the time of the special theory of relativity. Atomic clocks behave in the same way. Pendulum clocks do not automatically show the time of the SRT or ART. For them, the clock rate must be resynchronized with the display of one second. Due to the mechanical coupling, it is easier to change the pendulum speed with pendulum clocks. With atomic clocks, the number of caesium-133 oscillations for one second is adjusted with correction values. Of course, it then no longer shows the defined second as a second.

But pendulum clocks too depend on the gravitational field and go too imprecisely to measure what I intend to do. There should be a physically regular phenomenon, with a frequency and precision similar to atomic clocks, but independent of the gravitational field. I describe the possibilities I imagine in the following chapter. These clocks would also display the time in the sense of Einstein. He said, "Time is something you can tell from a watch." But it's not the time of special or general relativity.

Perhaps such a phenomenon has already been discovered, but then discarded again because it is believed that the clocks cannot run consistently. We want to call such watches K watches.

What time values would such a clock display? The measured values that can be achieved with such clocks are quite complicated, since the earth not only rotates, but also orbits the sun and this also around the centre of the galaxy, which is also moving. That is why I would first like to omit the effects that result from the earth's motion around the sun after the Lorentz transformations and also the gravitational effects of the earth. Let's **spin a ring in a void** with a speed and size equal to that of the equator and assume it wouldn't move sideways to the gravitational field here.

First, let's place 3 K clocks and 3 atomic clocks next to each other on this ring and synchronize them with each other. The atomic clocks are supposed to indicate the defined second and the K clocks are adjusted accordingly. The atomic clock A(**s**) stays together with the K-clock K(**s**). The atomic clock A(**w**) together with the K-clock K(**w**) is slowly moved along the ring in a westerly direction, and the atomic clock A(**e**) together with the K-clock K(**e**) in an easterly direction. We assume that the clocks are transported slowly, so that the effect of time dilation for the atomic clocks within the rotating ring plays only a minor role.

When the clocks come together again, clocks A(s), K(s), K(w) and K(e) show the same time. The clock A(w) is ahead by the amount m and the clock A(e) is behind by the amount m. The amount m depends only on the circumference of the ring and the tangential speed with which the ring rotates. It corresponds to the Sagnac effect.

## 5.6 Development of the formula for calculating m and n

In order to make it clear which value is measured by whom, those from the non-circulating ring are supplemented with $_R$ and those from the rotating ring are supplemented with $_K$.

The timing of the atomic clocks should correspond to the defined second.

The lengths are measured with atomic clocks in light seconds Ls. This has the advantage that the measured value for a distance and the time that a light signal needs to cover this distance only differ in the units of seconds and light seconds, but the amount is the same. If we only measure this with one clock, it is easy and for all observers a clear value that corresponds to half the time for the outward and return journey. If we only measure the easy way with two clocks, the value depends on the spatial synchronization of the clocks, which does not have to be the same for all observers.

For the outward path of a time signal, which should correspond to the direction of motion from the orbiting ring, the path relative to the stationary ring is denoted by A and relative to the orbiting ring by a. For the opposite direction correspondingly with B and b.

The circumference of the ring at rest is denoted by U and the circumference of the ring in motion by u.

$$U_R = (A_R + B_R) / 2$$

Since the lightning components reach the observer from both directions at the same time in the stationary ring is here:

$$U_R = A_R = B_R$$

Since the ring in motion should have the same spatial dimensions from the ring at rest:

$$U_R = u_R$$

Since length contraction occurs in the moving ring,

$$u_K = U_R / \text{square root}(1 - v^2 / c^2).$$

The tangential speed with which the circling ring moves is v and is given in c. According to the Lorentz transformations, two observers moving towards each other measure the speed of the other in the same way. So v does not need to be distinguished.

The time $a_R$ until the flash reaches the observer moving on the ring in the motion direction can be calculated as follows:

$$v * a_R + U_R = y \text{ and } c * a_R = y \qquad \text{So}$$

$$U_R = c * a_R - v * a_R = (c - v) * a_R \qquad \text{So}$$

$$\mathbf{a_R = U_R / (c - v)} \qquad \qquad (1)$$

For the opposite direction applies:

$$U_R = c * b_R + v * b_R = (c + v) * b_R \qquad \text{So}$$

$$\mathbf{b_R = U_R / (c + v)} \qquad \qquad (2)$$

The time difference $m_R$ measured from the stationary ring for the arrival of the lightning components at the circling observer is:

$$\mathbf{m_R = (a_R - b_R)} \qquad \qquad (3)$$

Allowing for time dilation, the time difference measured by the orbiting ring is:

$$\mathbf{m_K = m_R * square\ root(1 - v^2 / c^2)} \qquad \qquad (4)$$


This applies to the entire ring. To approach the length of the bar to be rotated, we first consider length contraction. The moving ring shortens, creating a gap. Measured from the ring at rest, the length of this part, denoted by bR, is only:

$$\mathbf{bR_R = U_R * square\ root(1 - v^2 / c^2)} \qquad \qquad (5)$$

Due to the resulting gap, the time signals can no longer return to the moving observer. So far we have used the ring in both directions, i.e. a double ring length. Now we split the ring just opposite the observer emitting the flash. Therefore we have to halve the times for the light propagation (* 0.5) and thus also the result $m_R$. Nothing else changes in principle.

For the values of this subsection, $n_R$ and $bR_R$, and the values of the whole ring, $m_R$ and $U_R$, the relationship is:

$$n_R / (m_R * 0.5) = bR_R / U_R \quad | \text{ in it we replace } bR_R \text{ formula (5)}$$

$$n_R / (m_R * 0.5) = U_R * square\ root(1 - v^2 / c^2) / U_R$$

$$| \text{ let's shorten } U_R \text{ and multiply by } (m_R * 0.5)$$

$$\mathbf{n_R = (m_R * 0.5) * square\ root(1 - v^2 / c^2)} \qquad \qquad (6)$$

If we take the time dilation into account, then the time difference $n_K$ measured in the moving section of the ring corresponds to:

$$n_K = n_R * square\ root(1 - v^2 / c^2) \mid \text{ we replace } n_R \text{ from formula (6)}$$

$$n_K = m_R * 0.5 * square\ root(1 - v^2 / c^2) * square\ root(1 - v^2 / c^2)$$

$$| \text{ we replace } m_R \text{ from formula (3)}$$

$$n_K = (a_R - b_R) * 0.5 * (1 - v^2 / c^2) \mid \text{ substitute } a_R \text{ and } b_R \text{ formula}$$

(1+2)

$$n_K = (U_R / (c - v) - U_R / (c + v)) * 0.5 * (1 - v^2 / c^2)$$

The Lorentz transformations show that a body continues to measure itself with the same length even after an acceleration. This also corresponds to the relativity principle of the STR. Thus the length of the accelerated ring section $bR_K$ is measured from the circling ring: $bR_K = U_R$.

Inserted into the formula:

$$\mathbf{n_K = (bR_K / (c - v) - bR_K / (c + v)) * 0.5 * (1 - v^2 / c^2)} \qquad (7)$$

These values are all only measured from the circling ring. The length of the bar $bR_K$ can be easily determined here and corresponds to half the time that a light signal needs for the outward journey plus the return journey in light seconds. The section can therefore be shortened to the desired length of a bar.

## 5.7 Measurement of the movement to the gravitational field

Let's first develop unambiguous events from the point of view of the stationary ring and see how the events unfold without contradiction when the bar is rotated on the circling ring.

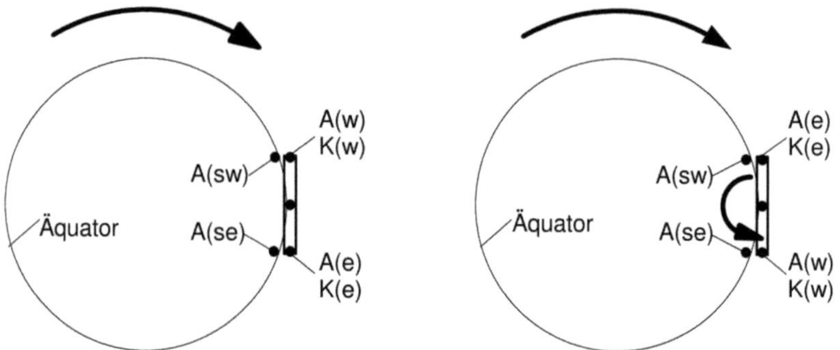


**Figure 13a and b:** A chariot or beam stands east-west at the equator and is rotated about its center. The equator is viewed from the South Pole. **A(sw)** and **A(se)** are the light/atomic clocks that are placed on the earth at the west or east end of the beam and remain there. **A(w)** and **A(e)** are the light/atomic clocks rotated with the bar. **K(w)** and **K(e)** should be clocks that are not under the influence of the gravitational field and are rotated with the bar.

Now we put clocks **A(w)** and **K(w)** together on the west end of a bar and clocks **A(e)** and **K(e)** on the east end. Clocks **A(sw)** and **A(se)** should be mounted on the ring at these locations and remain there when the bar is turned.

Clocks **A(w)** and **K(w)** are co-located at **A(sw)** and therefore indicate the same time, and their respective time signals 1 propagate with a uniform front. The same applies to the clocks **A(se)**, **A(e)** and **K(e)** located in the area of the other end of the bar.

The clocks should be spatially synchronized with UTC. The distance between W and O, measured with their clocks, should be one light second.

For example, the time displayed by a clock should be represented as **A(sw)**[1] for clock **A(sw)** at time 1. The time signal sent at that point in time as a(sw)[1]. The time signals sent out with them by the clocks located at their location are appended with a +.

So e.g. a(sw)[1] + a(w)[1] + k(w)[1]. The signal emitted by the west end clocks at UTC time 1.

The time signals should be reflected at the other end and sent back with the time signals from the clocks located there. The times of the clocks there are also appended with + in a new line.

If the clocks **A(se)** and **A(sw)** send out their time signal 1, then the time signal 1 sent in a westerly direction by clock **A(se)** arrives at clock **A(sw)** at 1 + a - n seconds. The eastbound time signal of clock **A(sw)** at clock **A(se)** at 1 + a + n seconds. The value for a and n are each measured in the proper time of a clock. The clear event at the west end of the bar looks like this when time signal 1 arrives:

```
A(sw)[1+a-n] + A(w)[1+a-n] + K(w)[1+a-n]
   a(se)[1]   +    a(e)[1]   +    k(e)[1]
```

The following list shows the sequence of the unambiguous events. The time values displayed by the clocks and the motions of the time light signals are calculated from the point of view of the non-rotating ring.

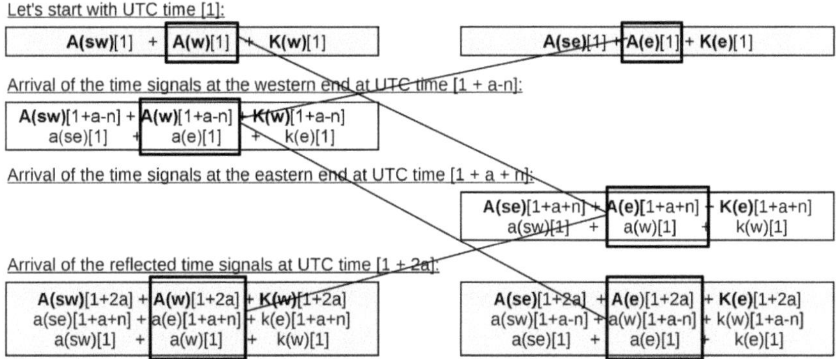


Measured from all clocks, the time signal a(o)[1] to the west takes (a − n) seconds and the time signal a(w)[1] to the east (a + n) seconds. The times for the way back are reversed accordingly.

Now let's flip the beam. Then the clocks **K(w)** and **K(e)** are still synchronous with the clocks **A(sw)** and **A(se)**, which represent UTC. However, clock **A(w)** is n seconds behind UTC because it has slowed down on its way east, and clock **A(e)** is n seconds ahead of UTC because it has gone faster on its way west. Turning should take d seconds.

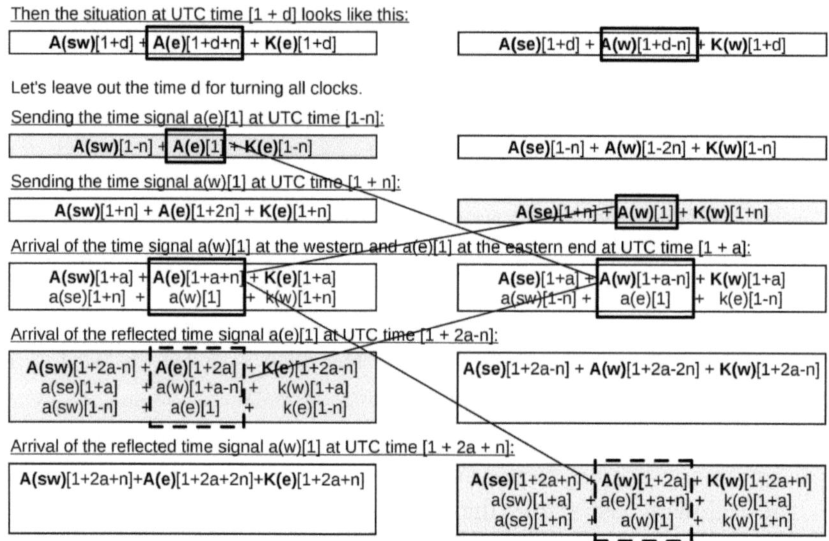


Let's just consider the pair of clocks A(w) and A(e). Before turning over, the time signal [1] of clock A(e), i.e. a(e)[1], arrives at A(w) at event **A(w)**[1+a-n] and not just halfway from 2a, which would correspond to **A(w)**[1+a]. The time signal [1] of clock A(w), i.e. a(w)[1], only arrives at event **A(e)**[1+a+n] at clock A(e) and not already at event **A(e)**[1+n].

Due to the spatial synchronization of clocks A(w) and A(e) according to UTC, a longer time is measured for the signal from A(w) to A(e) than vice versa.

Flipping it doesn't change anything. Furthermore, the time signal a(e) [1] arrives at **A(w)**[1+a-n] and the time signal a(w)[1] arrives at **A(e)** [1+a+n]. The time signals arrive at the clocks unchanged. The clocks now measure a longer time for the west direction than for the east direction, which is causally wrong. It only comes about because the clocks A(w) and A(e) are now no longer synchronous with UTC. If two lasers (lasers behave like light clocks) send their beams to each other, there would be no shift in the interference pattern during the turning.

In order to spatially synchronize the clocks A(w) and A(e) according to Einstein's definition of simultaneity, so that the time used by the time signals is measured equally for both directions, we can, for example, adjust the clock A(e) by the amount n reset.

Then the start events at UTC time [1], at which A(w) also transmits the time signal a(w)[1], look like this:

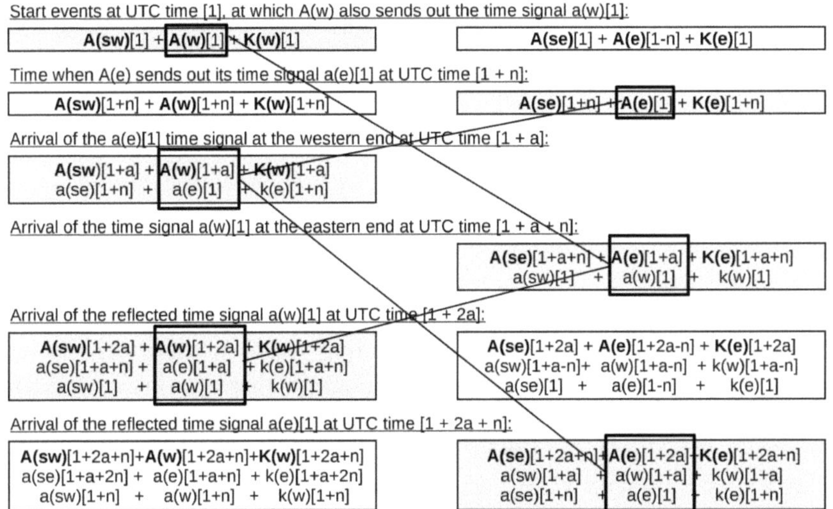


Again, consider only the clocks A(w) and A(e). With this spatial synchronization of the clocks, they measure the same value for the duration that the time signals need both in the east and in the west direction, namely half of 2a, i.e. a.

If we now turn the bar around, which should take d seconds.



Clock A(e) has rated n seconds faster on its way west. It is now back in sync with UTC after having previously lagged behind UTC by n seconds. Clock A(w) has rated slower by n seconds on its way east and is now n seconds behind UTC. From this situation, the sequence looks the same as in the previous example, only with the roles of A(w) and A(e) reversed. And here, too, the arrival of the time signals does not change when the bar is turned around.

For all observers, the signals for the outward and return journey together take 2a seconds. From the spatial synchronization of the clocks

A(w) and A(e) according to Einstein's definition of simultaneity, the same time is measured for the outward and return journey, which does not change when the bar is rotated. In the case of rotation, however, this spatial simultaneity is not compatible with the observations repeated in the sky, the time signals sent over the rest of the ring or the clock transport over the rest of the ring.

With this, one can causally refute the spatial synchronization according to Einstein's definition of simultaneity as incorrect for rotation processes by means of clear measurement results.

If one had the possibility of passing on information at a speed higher than the speed of light, one could narrow the spatial simultaneity more than with light signals. With that, Einstein's definition of simultaneity would also have to be causally refuted as wrong for linear motion. Here in the ring system, however, it would only confirm spatial synchronization according to UTC. (As far as it is a ring resting to the gravitational field.)

If one considers the postulates of the special theory of relativity to be actually given in the universe surrounding us, then there can be no higher speed than the speed of light. Of course there can be no WARP drive or wormholes. At least no wormholes leading back into our universe. Whether I send information faster than light or transport it with a spaceship with WARP drive makes no causal difference.

Let's consider the situation where all clocks are in sync in UTC. After flipping the bar, clock A(e) is not only ahead of UTC by $n_K$ seconds, but also of clock K(e). And clock A(w) is not only $n_K$ seconds behind UTC, but also behind of clock K(w). The time difference $n_K$ that develops between K(w) and A(w) depends only on the distance they cover in the east-west direction and the tangential speed with which the ring rotates.

What are the results if this ring is now mounted on a larger ring C, which corresponds to the earth's orbit around the sun. The UTC clocks are also spatially synchronized on this ring. Here we want to call the time $UTC_S$. The time of the clocks stationary on the small ring B $UTC_E$.

Logically, the clocks on ring B behave like the clocks on the bar in the previous example, which you just keep turning here. Here, too, the clocks in the $UTC_E$ run slower when moving in the direction of earth motion and faster in the opposite direction compared to the clocks in the $UTC_S$. However, as with the bar, this has no influence on the arrival of the time signals within the ring B, as long as the times of transmission and reception are measured with light clocks or atomic clocks.

However, since the K-clocks do not behave in accordance with the Lorentz transformations, a time difference with a sinusoidal change occurs between the atomic clocks and the K-clocks standing next to them. Instead, they show a behaviour corresponding to the bar example, i.e. no change to the $UTC_S$ clocks.

However, the earth orbits with the sun around the centre of the Milky Way. If our example is extended by a corresponding effect, then there is a sinusoidally fluctuating time difference between the K clocks and the $UTC_S$. This would overlay the curve of the time difference between the $UTC_E$ and the K clocks. We would therefore only have a comparable curve for the fluctuation of the time difference after a year at the earliest.

## 5.8 No difference in measured values achievable with atomic clocks between rotation and inertial systems

Let's look at a section of a cylinder surface (cylinder diameter three times the diameter of the Milky Way) with an edge length of one astronomical unit. In this section, the clocks are spatially synchronized according to Einstein's definition of simultaneity. Under these conditions, the clock rate and the propagation of light signals are the same as in an inertial system. If we let two or more cylinder surfaces move past each other and only look at the atomic clocks and time light signals, then the measurement results and calculations correspond exactly to the conditions of inertial systems. But we could still send signals along the cylinder, even if it would take a few million years. With this we could generate measurements that would causally represent the spatial synchronization of the clocks as wrong according to Einstein's definition of simultaneity.

Only when we detach this surface from the cylinder do we actually have a straight-line motion, even if this hardly differs from the other motion in a measurable way. This also eliminates the centrifugal force. It is also questionable whether a time signal that we send through the universe in one direction would reach us again from the other direction. But we can still watch the universe pass us by. However, there are no observations in the sky that are repeated.

Who can take the big step with me and question the absolute constancy of the speed of light and assume it is constant to the gravitational field?

Who can still assume that the gravitational field, formed from the superimposed fields of the mass particles, would be a plastically deformable field in its effect? Then the Mercury anomaly and the Schiff effect could not only be superimpositions of the gravitational field, as Einstein interpreted it, but plastic changes of the field. This would also mean that the earth slightly distorts and entrains the gravitational field as it moves around the sun, causing the fly-by anomaly. The effect could be even greater if the masses that locally determine the gravitational field move in a common direction. The gravitational field causes the centrifugal force. However, if this rotates a bit with rotating galaxies, it looks from the outside as if the stars are moving too quickly. But it is exactly the right speed for the prevailing gravitational field, even without dark matter.

I would consider that the simpler solution and therefore the right one according to Ockham's razor.

Which mathematician with access to the astronomical data would like to test this solution?

### *5.9 Basics for the watch you are looking for.*

According to Einstein, according to the principle of relativity, it is impossible to determine by mechanical experiment whether we are at rest or in uniform motion.[26]

As I already wrote about space in chapter 2.1, the question about absolute space is like the question about God. The human being as an integrated part in the universe surrounding us cannot measure this. Man can only make comparisons and these comparisons are usually linked to assumptions.

So we can only specify a rest for something that we can also measure. For example, we can determine through mechanical experiments whether we are at rest to a sound transport medium. With that we would have determined a rest. But this is not what is meant in the theory of relativity. It is about a rest to what determines the speed of light.

If there is nothing that determines this, apart from a principle, then of course it cannot be measured either. But in what do gravitational waves

---

26   [9] S.28; [21] S.16

travel. If there is such a field, it does not immediately mean that we can also measure the motion against it. Just because we haven't invented radio yet doesn't mean radio waves don't exist.

With rotation, we can clearly state that there is something to which the speed of light is constant and to which the observer can also move, therefore here the speed of light is not constant to the observer either.

However, two rotating circles can also move in a straight line to each other. Also for this motion we could construct a much larger ring, but it would not be a fundamental solution for the rectilinear motion.

However, the arguments presented here in the book should arouse the willingness to imagine that the principle of motion can be applied to the gravitational field. Then there should also be a willingness to look for an instrument with which this motion can be measured.

As I wrote about radio waves in chapter 3.9, I believe that the electromagnetic field has nothing to do with the gravitational field. The charged particles like electron and proton are also mass particles, in that sense the fields are already connected with each other. But you can't make neutral mass particles vibrate with radio waves. It might be possible to use this to find a clock generator that would have the same clock frequency on the mountain as it did in the valley. This would of course lead to time differences to an atomic clock over the course of the day and year.

But even if you compare these clocks directly with each other, they would fluctuate sinusoidally in their time relative to each other. But that would be because, from their point of view, the time signals for the single path need different times. This is depending on the alignment of the axis between them in respect to the direction of movement against the gravitational field and the speed relative against the gravitational field.

Length contraction also changes the distance between these clocks as they rotate. As a result, there is no change in the propagation time of a time signal for the total for the outward and return journey during the rotation. This is only changed by changing the speed to the gravitational field. Because there is no time dilation with these clocks, they measure a change in transit times, but not light clocks.

How about the radioactivity? The radioactive isotope chlorine-36 has the greatest rate of decay around the turn of the year, when the distance to the sun is shortest. The density of neutrinos originating from the sun is also greatest here. Does this have something to do with each other causally or is it just a coincidental effect, like with the storks and the births at the beginning of the twentieth century.

However, this is also the time when the earth is moving towards the microwave background radiation (CMB) the fastest in the course of the year. I had already described above that the gravitational field can also move against the CMB, i.e. the speeds relative to the gravitational field and the CMB do not have to match in direction and size, but roughly they should match.

When the earth moves fastest towards the gravitational field at the turn of the year, the atomic clocks also go slowest here. The radioactivity then seems to increase relative to the clocks.

This should then also affect all radioactive processes to the same extent, insofar as they are not influenced by the gravitational field. However, several effects could also overlap here.

An easy way to check that would be to let a satellite with an atomic clock and a corresponding radioactive sample orbit the earth. The orbit should be as far as possible from the earth but remain the same and the axis should point perpendicularly to the sun. At new moon or full moon the gravitational changes on the orbit would be small and the neutrino influence would be the same on the hole orbit. The speed to the CMB and also the gravitational field should then change depending on the time of year and the direction of the orbit. Of course, the satellite can also move on any other orbit.

There is certainly already a lot of data on changes in radioactivity. On the basis of which one could check whether they support this thesis or make it appear rather improbable. In any case, I don't think finding such a clock would be as costly as searching for dark matter.

However, there are already observations of the exact measurement of the microwave background radiation. This is where anomalies like the "axis of evil" and others that correlate with the ecliptic occur. Perhaps these can be eliminated if they are related to a sinusoidal course of the atomic clocks and thus variable measured values.

Since the gravitational field can also move to the microwave background radiation, an indication could also be gained as to whether the gravitational field moves along with the rotation around the center of the Milky Way, so that the solar system does not move too quickly around the center of the Milky Way.

# Comment on plagiarism

This book is written in an unconventional way. It is also intended to break with conventions, as this too often involves driving along tracks that have already been laid. With any framework, you can no longer see things outside of that framework.

My citation style is certainly not entirely correct, but the content is probably unmistakably comprehensible. I would have liked to have made more references to passages from the books I've read. I usually can't find these places any more. That's why I added the book "Theory of Relativity for Dummies". I think it will not be dismissed as nonsense by technically competent physicists. Therefore, the references mentioned here should also be universally valid. They can also be found in the other books.

Quotations I wanted are also made clear. Some texts could also come almost verbatim from other books. I didn't consciously suppress that as a quote, I just can't remember reading it that way in the other books.

I read the book playful thinking by E.de Bono[5] when I was about 13 years old. 40 years later I wanted to quote from it and then read a lot of it again. I was shocked that some of the ideas that I considered to be my very own thoughts were already written here. Some of my formulations are almost word for word exactly the same. So if some of my formulations should already be in this way somewhere else, that only speaks for the quality that they have made an impression on me. So if someone finds such a passage, I would be happy about a hint, then I can also report it here as a quote as a convincing formulation. I haven't read my core statements like that in other books. Here I would be more interested in competent physicists or philosophers dealing with it critically. But I would also be happy to hear from anyone else. The easiest way is via <u>buch@darmer.de</u>

# literature

[1]   Internet pages from Wikipedia

[2]   Springer, V.: *Die Entstehung der Galaxien*, Physik Journal 2 (2003) Nr.6 S. 31

[3]   Bauer, M.: *Vermessung und Ortung mit Satelliten*, 2. Auflage Wichmann Verlag 1992

[4]   Bergmann, Schaefer: Lehrbuch der Experimentalphysik, Band 1: *28.1 Die Axiome der speziellen Relativitätstheorie*, 11. Aufl. de Gruyter Verlag, 1998

[5]   de Bono, E.: *Das spielerische Denken*, Scherz Verlag, 1967
       de Bono, E.: *In 15 Tagen Denken lernen*, Rowohlt Verlag 1967

[6]   Born, W.: *Die Relativitätstheorie Einsteins*, Unveränderter Nachdruck 5. Auflage Springer-Verlag 1969

[7]   Cremer, T.: *Interpretationsprobleme der speziellen Relativitätstheorie*, 2. Auflage 1990 Verlag Harri Deutsch

[8]   Einstein, A.: *Zur Elektrodynamik bewegter Körper*, Annalen der Physik **17**, Seiten 891-921,1905

[9]   Einstein, A.: *Grundzüge der Relativitätstheorie*, 5. Auflage Vieweg Verlag 1984

[10] Einstein, A.: *Über die spezielle und die allgemeine Relativitätstheorie*, 23. Auflage Vieweg Verlag, 1988

[11] Goenner, H.: *Spezielle Relativitätstheorie*, Elsevier GmbH, Spektrum Akademischer Verlag, 2004

[12] Hetznecker, H.: *Relativitätstheorie für Dummies* ePub 1. Auflage 2018 WILEY-VCH Verlag

[13] Hossenfelder, S.: *Das hässliche Universum*, E-BOOKS S.Fischer Verlag 2021

[14] Hoffmann, B.: *Einsteins Ideen Das Relativitätsprinzip und seine historischen Wurzeln*, 1997 Spektrum Akademischer Verlag

[15] Lämmerzahl, C.: *Detektivarbeit krönt Langzeitprojekt*, Physik Journal 10 (2011) Nr. 7 Seite 19

[16] Lämmerzahl, C.: *Einstein besser bestätigt*, Physik Journal 2 (2003) Nr.12 Seite 18

[17] Marder, L.: *Reisen durch die Raum-Zeit*, Vieweg Verlag, Nachdruck 1982

[18] Mücklich, Dr. A.: *Das verständliche Universum*, e-book Books on Demand GmbH, Norderstedt

[19] Neffe, J.: *Der Geistesmächtige* Der Spiegel Nr.50/1999 S. 271

[20] Rievers, B., Lämmerzahl, C.: *Pioneer-Anomalie entschlüsselt*, Annalen der Physik **523**, 439 (2011)

[21] Rindler, W.: *Relativitätstheorie Speziell, Allgemein und Kosmologisch*, Deutschausgabe 2016 WILEY-VCH Verlag

[22] Ruder, H. u. M.: *Die Spezielle Relativitätstheorie*, Vieweg Verlag, 1993

[23] Schröder, U. E.: *Spezielle Relativitätstheorie*, Verlag Harri Deutsch 1987

[24] Sexl, R., Schmidt, H.K.: *Raum – Zeit – Relativität*, 3. Auflage Vieweg Verlag, 1991

[25] Weigert, A., Wendker, H.J., Wisotzki, L.: *Astronomie und Astrophysik*, 5. Auflage Wiley-VCH, 2011